Addressing Ethnic Conflict through
Peace Education

# Addressing Ethnic Conflict through Peace Education

*International Perspectives*

Editors

*Zvi Bekerman*
*Claire McGlynn*

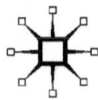

ADDRESSING ETHNIC CONFLICT THROUGH PEACE EDUCATION
© Zvi Bekerman and Claire McGlynn, 2007.

All rights reserved. No part of this book may be used or reproduced in any manner whatsoever without written permission except in the case of brief quotations embodied in critical articles or reviews.

First published in 2007 by
PALGRAVE MACMILLAN™
175 Fifth Avenue, New York, N.Y. 10010 and
Houndmills, Basingstoke, Hampshire, England RG21 6XS
Companies and representatives throughout the world.

PALGRAVE MACMILLAN is the global academic imprint of the Palgrave Macmillan division of St. Martin's Press, LLC and of Palgrave Macmillan Ltd. Macmillan® is a registered trademark in the United States, United Kingdom and other countries. Palgrave is a registered trademark in the European Union and other countries.

ISBN 978-1-349-53311-4        ISBN 978-0-230-60358-5 (eBook)
DOI 10.1057/9780230603585

Library of Congress Cataloging-in-Publication Data

　　Addressing ethnic conflict through peace education : international perspectives / edited by Zvi Bekerman and Claire McGlynn.
　　　p. cm.
　　Includes bibliographical references and index.
　　　1. Peace—Study and teaching—Case studies. 2. Ethnic conflict—Case studies. 3. Conflict management—Case studies. I. Bekerman, Zvi.
II. McGlynn, Claire.
JZ5534.A33 2007
303.6'6—dc22                                                            2006050735

A catalogue record for this book is available from the British Library.

Design by Newgen Imaging Systems (P) Ltd., Chennai, India.

First edition: April 2007

10 9 8 7 6 5 4 3 2 1

Transferred to Digital Printing in 2013

# Contents

| | |
|---|---|
| *List of Figures and Tables* | ix |
| *Acknowledgments* | xi |
| *List of Contributors* | xiii |
| Introduction<br>Zvi Bekerman and Claire McGlynn | 1 |

### Section I  Systemic Issues

| | | |
|---|---|---|
| One | Desegregation and Resegregation: The Legacy of Brown versus Board of Education, 1954<br>Tony Gallagher | 9 |
| Two | Moving from Piecemeal to Systemic Approaches to Peace Education in Divided Societies: Comparative Efforts in Northern Ireland and Cyprus<br>Laurie Shepherd Johnson | 21 |
| Three | Multiple Realities and the Role of Peace Education in Deep-Rooted Conflicts: The Case of Cyprus<br>Maria Hadjipavlou | 35 |
| Four | Reconciliation and Peace in Education in South Africa: The Constitutional Framework and Practical Manifestation in School Education<br>Elmene Bray and Rika Joubert | 49 |

### Section II  Teachers and Students

| | | |
|---|---|---|
| Five | Color Coded: How Well Do Students of Different Race Groups Interact in South African Schools?<br>Saloshna Vandeyar and Heidi Esakov | 63 |

| | | |
|---|---|---|
| Six | Challenges in Integrated Education in Northern Ireland<br>*Claire McGlynn* | 77 |
| Seven | Developing Palestinian-Jewish Bilingual Integrated Education in Israel: Opportunities and Challenges for Peace Education in Conflict Societies<br>*Zvi Bekerman* | 91 |
| Eight | Peace Education in a Bilingual and Biethnic School for Palestinians and Jews in Israel: Lessons and Challenges<br>*Ilham Nasser and Mohammed Abu-Nimer* | 107 |
| Nine | Is the Policy Sufficient? An Exploration of Integrated Education in Northern Ireland and Bilingual/Binational Education in Israel<br>*Joanne Hughes and Caitlin Donnelly* | 121 |

### Section III  Curriculum and Pedagogy

| | | |
|---|---|---|
| Ten | Education for Peace: The Pedagogy of Civilization<br>*H.B. Danesh* | 137 |
| Eleven | Learning to Do Integrated Education: "Visible" and "Invisible" Pedagogy in Northern Ireland's Integrated Schools<br>*Chris Moffat* | 161 |
| Twelve | From War to Peace: An Analysis of Peace Efforts in the Ife/Modakeke Community of Nigeria<br>*Francisca Aladejana* | 173 |

### Section IV  Adult Education

| | | |
|---|---|---|
| Thirteen | Toward Sustainable Peace Education: Theoretical and Methodological Frameworks of a Program in South Africa<br>*Tim Houghton and Vaughn John* | 187 |
| Fourteen | Learning and Unlearning on the Road to Peace: Adult Education and Community Relations in Northern Ireland<br>*Paul Nolan* | 201 |

### Section V  Teacher Education

| | | |
|---|---|---|
| Fifteen | The Reconstruction of the Teacher's Psyche in Rwanda: The Theory and Practice of Peace Education at Kigali Institute of Education<br>*George K. Njoroge* | 215 |

| Sixteen | Post-Soviet Reconstruction in Ukraine: Education for Social Cohesion<br>*Tetyana Koshmanova and Gunilla Holm* | 231 |
| Seventeen | Teacher Preparation for Peace-Building in United States of America and Northern Ireland<br>*Candice C. Carter* | 245 |
| *Index* | | 259 |

# List of Figures and Tables

**Figures**

| | | |
|---|---|---|
| 13.1 | CAE Peace Education Program | 191 |
| 17.1 | Continuum of Education Approaches to Cross-Cultural Peace Building | 254 |

**Tables**

| | | |
|---|---|---|
| 5.1 | Profile of Schools | 66 |
| 8.1 | Language Fluency | 115 |
| 12.1 | Percentage Response on Assessment of Teaching Strategies | 178 |
| 12.2 | Imbibing the Culture of Peace from Social Studies | 180 |
| 17.1 | NAME Criteria for Evaluating Curriculum Standards | 249 |

# Acknowledgments

We want to express our deep sense of appreciation, first and foremost to Alan Slifka and the Alan B. Slifka Foundation (United States of America) for their substantive support throughout all stages of the initiative, which allowed the group of scholars represented in this volume to come together and discuss their perspectives on the multiple and complex issues that burden sustained educational efforts toward peace in areas of protracted conflict. On a more personal note, we wish to thank Sarah Silver, the Slifka Foundation director, for her guidance and support in planning the conference and helping us focus on the needs of the field in all that concerns peace and coexistence education.

Thanks and appreciation are also much in place to the Melton Centre for Jewish Education, the Truman Institute for the Advancement of Peace, and the Swiss Center for Conflict Resolution, all located at the Hebrew University (HU) in Jerusalem. We are also grateful to Stranmillis University College and the School of Education, Queen's University, Belfast for their continuous support and encouragement in bringing this initiative to fruition. Again, specific individuals associated with these institutions deserve our recognition. Dr. Howard Deitcher, Professor Eyal Ben-Ari, Professor Yaakov Bar-Siman-Tov (respectively from the HU), Professor Richard McMinn and Dr. Les Caul of Stranmillis University College and Professor Tony Gallagher of the School of Education, Queen's University. We also wish to thank Julia Schlam-Salman, an outstanding graduate student at the HU. She had a leading role in the coordination of this project beginning with the conference and concluding with the publication of this volume. Coordinating is undoubtedly one of Julia's strengths but mostly we benefited from her critical and insightful mind that, throughout, challenged and bettered our work. Finally, we owe thanks and recognition to Professor Alan Smith from the UNESCO Centre, University of Ulster and his matchmaking skills that enabled us to ultimately became friends and coinitiators/conveyers/authors of this challenging and fruitful initiative.

Zvi Bekerman and Claire McGlynn

# List of Contributors

**Francisca Aladejana**
*Obafemi Awolowo University (Ile-Ife, Nigeria)*
Dr. Aladejana is a senior lecturer in the Faculty of Education, Obafemi Awolowo University, Ile-Ife, Nigeria. Her area of specialization and research interests include science education, gender studies, and early childhood education and peace education. She is currently the acting Director of the Institute of Education, Obafemi Awolowo University, Ile-Ife.

**Mohammed Abu-Nimer**
*American University (Washington DC, United States)*
Dr. Abu-Nimer is an associate professor at the American University's School of International Service in International Peace and Conflict Resolution and Director of the Peacebuilding and Development Institute. He is an expert on conflict resolution and dialogue for peace and cofounder/coeditor of the Journal *Peacebuilding and Development*. Recent publications include: *Reconciliation, Coexistence, and Justice in Interethnic Conflicts* (2001) and *Peacebuilding and Nonviolence in Islamic Context: Theory and Practice* (2003).

**Zvi Bekerman**
*The Hebrew University (Jerusalem, Israel)*
Bekerman teaches anthropology of education at the School of Education and the Melton Center, Hebrew University of Jerusalem. He is also a Research Fellow at the Truman Institute for the Advancement of Peace, Hebrew University. His main interests are in the study of cultural, ethnic, and national identity, including identity processes and negotiation during intercultural encounters and in formal/informal learning contexts. Since 1999 he has been conducting a long-term ethnographic research project in the integrated/bilingual Palestinian-Jewish schools in Israel. He has also recently become involved in the study of identity construction and development in educational computer-mediated environments. He is the coeditor (with Seonaigh Macpherson of *Diasora, Indigenous, and Minority Education: A International Journal* (Lea, 2007) and his most recent book is *Learning in Places: The Informal Educational Reader* (Peter Lang, 2006) (edited together with Nicholas Burbules and Diana Keller Silverman).

**Elmene Bray**
*University of South Africa (Pretoria, South Africa)*
Dr. Bray is a professor of law in the Department of Constitutional, International, and Indigenous Law, University of South Africa and deputy director of the Interuniversity Centre for Education Law and Policy (CELP). She teaches education law and is author and coauthor of several education law books. She has published widely in both national and international journals.

**Candice C. Carter**
*University of North Florida (Jacksonville, United States)*
Dr. Carter is an associate professor at the University of North Florida. Her interests include domestic and international teacher preparation, citizenship education, social studies instruction, peace curriculum, arts-based instruction including peace through arts and peace literature. Dr. Carter has the website http://peacemaker.st. with information about peace issues, events and resources.

**H.B. Danesh**
*International Education for Peace Institute (Vancouver, Canada)*
Professor H.B. Danesh is founder and director of the International Education for Peace Institute. He devised the Education for Peace Program that was first piloted in Bosnia and Herzegovina. He is a psychiatrist and professor emeritus of Conflict Resolution and Peace Studies at Landegg International University.

**Caitlin Donnelly**
*Queen's University (Belfast, Northern Ireland)*
Dr. Donnelly is a lecturer in education at the School of Education, Queen's University Belfast. She has conducted research on school governance, school ethos, and education policy in Northern Ireland. She has published in a range of peer-reviewed journals and is lead editor and contributor to *Devolution and Pluralism in Education in Northern Ireland* (Manchester University Press, 2006) Manchester University Press.

**Tony Gallagher**
*Queen's University (Belfast, Northern Ireland)*
Dr. Gallagher is a professor at Queens University Belfast and Head of the School of Education. His main research interest lies in the role of education in societies experiencing ethnic or religious conflict. He has published widely on these issues and his most recent book is entitled *Education in Divided Societies* (Palgrave/Macmillan, 2004).

**Maria Hadjipavlou**
*University of Cyprus (Nicosia, Cyprus)*
Dr. Hadjipavlou is an assistant professor in the Department of Social and Political Sciences, University of Cyprus. Her research interests include conflict resolution and international conflicts, gender and politics, and feminist approaches and theories. She has spent over 28 years working for and promoting peace building activities across the divide in Cyprus.

**Gunilla Holm**
*Western Michigan University (Kalamazoo, United States)*
Dr. Holm's area of specialization is Sociology of Education. Her research foci include issues related to race, ethnicity, class, and gender. She has published in both national

and international journals. She is vice president for North America for the International Sociological Association Research Committee on Youth.

**Tim Houghton**
*University of KwaZulu-Natal (Pietermaritzburg, South Africa)*
Dr. Houghton works for the Centre for Adult Education (CAE), University of KwaZulu-Natal, Pietermaritzburg as a researcher, curriculum developer, and materials writer, program and course coordinator, and teacher/facilitator. He currently coordinates the Peace Education Program within CAE and teaches peace education and peace studies. His interests include the role of conflict transformation within community education and development projects.

**Joanne Hughes**
*University of Ulster (County Antrim, Northern Ireland)*
Dr. Hughes is a professor in the School of Policy Studies, University of Ulster. Her main research interests include intergroup relations and community development in divided societies. Current research projects include an analysis of direct and indirect cross-community contact and tolerance in mixed and segregated areas of Belfast.

**Vaughn John**
*University of KwaZulu-Natal (Pietermaritzburg, South Africa)*
Dr. John is a trained psychologist who works as an adult educator at the University of KwaZulu-Natal, South Africa. His academic interests lie in teaching peace education and research methodology, and in researching learner support in open learning. He is currently director of the Centre for Adult Education (CAE) at the Pietermaritzburg campus of the University of KwaZulu-Natal.

**Laurie Shepherd Johnson**
*Hofstra University (Hemsptead, United States)*
Dr. Shepherd Johnson is professor and director of the Graduate Programs in Counseling in the School of Education and Allied Human Services at Hofstra University, New York. Her main research and training interests are in intercultural and systemic approaches to conflict transformation in schools and community.

**Rika Joubert**
*University of Pretoria (Pretoria, South Africa)*
Dr. Joubert is a senior lecturer in the Department of Education Management and Policy Studies at the University of Pretoria and the director of the Interuniversity Centre for Education Law and Policy (CELP). She is a coauthor of two books on Education Law and has published widely on education law issues and training in multicultural situations.

**Tetyana Koshmanova**
*Ivan Franko National University of Lviv (Lviv, Ukraine) and Western Michigan University (Kalamazoo, United States)*
Dr. Koshmanova's area of specialization is teacher education. She teaches teacher education courses at the Ivan Franko National University of L'viv and at Western Michigan University. Her research focuses on issues of culture and community in learning to teach, peace education, and Vygotskian perspectives in teacher education.

## Claire McGlynn
*Queen's University (Belfast, Northern Ireland)*
McGlynn is a lecturer in Continuing Professional Development at the School of Education, Queen's University, Belfast. Her main interests lie in gaining understanding of the impact of integrated (Catholic-Protestant) education in Northern Ireland, particularly with regard to the impact on social identity and on respect for diversity. As a teacher educator and researcher, McGlynn is committed to finding innovative methods of developing teacher education for diverse societies.

## Chris Moffat
*The Fortnight Educational Trust (Belfast, Northern Ireland)*
Chris Moffat edited the first major survey of integrated education in Ireland in 1993. She writes on education policy in Northern Ireland and education issues more generally. Recent work includes the Education Rights chapter in the "CAJ Handbook on Civil Liberties in Northern Ireland" (Northern Ireland: Committee for the Administration of Justice, 2003).

## George K. Njoroge
*Kigali Institute of Education (Kigali, Rwanda)*
Dr. Njoroge is an associate professor of Philosophy of Education at Kigali Institute of Education. His interests include education and genocide in Rwanda, lifeskills education, philosophy and HIV/AIDS, philosophical palavering for creative living, identity construction, citizenship education, and education and democracy. He is the coauthor of the book *The Democratization Process in Africa* (Nairobi: Quest and Insight Publishers, 2001) and coeditor of the book *Education, Gender and Democracy in Kenya* (Nairobi: Quest and Insight Publishers, 2001).

## Ilham Nasser
*George Mason University (Fairfax, United States)*
Dr. Nasser is an assistant professor in Early Childhood Education at George Mason University. She has spent over 20 years teaching and researching education and child development issues in both the United States of America and the Middle East. Recently, she has been focusing on bilingual education as a means for promoting peace education in early childhood settings.

## Paul Nolan
*Queen's University (Belfast, Northern Ireland)*
Paul Nolan is director of the Institute of Lifelong learning at Queen's University, Belfast. Prior to this appointment, he was director of the Workers' Educational Association in Northern Ireland and has been involved in many initiatives in political education.

## Saloshna Vandeyar
*University of Pretoria (Pretoria, South Africa)*
Dr. Vandeyar is a senior lecturer in the Department of Curriculum Studies in the Faculty of Education at the University of Pretoria, South Africa. She specializes in education and diversity, teacher professionalism, and ssessment strategies and practices.

## Heidi Esakov
Heidi Esakov is a Master of Education (M.Ed.) student at the University of Pretoria, South Africa under the supervision of Dr. Vandeyar.

# Introduction

## Zvi Bekerman and Claire McGlynn

We have both been involved for years now in the practical implementation and theoretical and empirical research of peace education in our respective countries, Northern Ireland and Israel. Alongside many, we have watched in recent decades the worldwide rise of ethnic tensions and sectarianism followed by bloody conflicts that, at times, seemed able to reach nothing but deadlock. With others, we share a faith in education as a possible solution to ethnic conflict. We want to emphasize the possible as we are not naive. We realize that institutionalized educational efforts can have and have had multiple uses, at times encouraging hatred and prejudice as much as, at times, actively becoming involved in trying to prevent them. We strongly believe education has a task, an important task to play in negotiating our views regarding that which is human and in need of dignity and recognition but we also know that education, all by itself, cannot achieve these goals. We see sustained educational efforts toward peace as a necessary, but not sufficient, condition that when unsupported by structural (visible political-economical-social) change might waiver. We maintain that peace education needs not only to struggle against dysfunctional human relationships but must also commit itself to more critical approaches through which to disclose the historical forces and political structures that generate and sustain conflict in our world.

We are well aware of the varied kinds of peace education, conflict resolution, education for democracy, multicultural education, and diversity training programs that are widespread all over the world, from Northern Ireland to Rwanda, from the United States of America to Germany, and from Israel to the former Yugoslavia. Such programs include school-based Peace Education curricula or programs to curb school violence, community-based meetings, seminars, youth summer camps, cooperative projects, and more. We know as well that, for the most part, these efforts are partial within the contexts of their implementation, specific curricular units within the school curricula, or short-term retreat encounters between the conflicting groups and that fewer are the efforts that approach peace and conflict resolution through long-term educational efforts. Also we know that peace educational work is, for the most part, substantiated and investigated through a small number of theoretical perspectives many of which rest on psychological understandings and that despite the wealth of short-term peace programs and projects, there is a great paucity of scholarly work to accompany it, thus making research on sustained educational efforts in this field even more rare. Needless to add, there are few comparative studies at hand that can help theoreticians sharpen their perspectives through interdisciplinary approaches and/or support educators involved in such initiatives in their complex task.

With these considerations in mind, three years ago we started a rather small initiative of which the present volume is its product. We have brought together in dialogue scholars representing a variety of disciplinary fields and methodological approaches, exemplifying the multiple venues of peace educational labor with an emphasis on long-term educational efforts. The present volume is the first international effort to present studies on sustained education for peace, coexistence, and reconciliation, particularly in high-conflict societies or societies making the transition from protracted conflict to the consolidation of peace. It presents the most recent theoretical advances in the field, analyzes the social, cultural, political, historical, and economical contexts within which peace education develops and must be critiqued, probes into the epistemology that nourishes its development and the practices that characterize its implementation, and reflects on the variety of educational contexts in which it is practiced.

Our efforts are geared toward academics, practitioners, and policymakers. The chapters consider prominent issues connected to integrated education, such as the impossibility of forcing integration in contested societies, the impact of educational integration on cultural/ethnic/religious identities, effective methods and strategies in integrated classroom work, and integrated education not as an end in itself but as a model for educational excellence in general and civic/democratic education in particular. Moreover, the chapters suggest practical educational policies regarding the roles of the key actors and sectors (government agencies/agents, parents, teachers, etc.) in promoting and developing tools and coherent approaches to sustained peace educational efforts and integrated education in conflict-ridden areas.

The dialogue we sustain raises multiple issues and not all are covered in this volume. One of the central issues that has become apparent is the need for a strong and consistent interdisciplinary approach to peace education. We are all in one way or another surprised at our relative ignorance regarding the multiple paradigmatical, theoretical, curricular, and practical perspectives from which our colleagues have come to confront issues related to peace/antiprejudice/coexistence/civic (or any other labeling our restricted contexts call or allow for when dealing with these issues) education. We have become aware once again of the dangers of academic compartmentalization, whilst justified by analytical needs but always reflecting issues of turf in academic areas of influence. Needless to say, all of us who believe that the educational field in general and peace education in particular necessitates a strict congruence between theoretical perspectives and practical activity, need to set turf aside. As for analytical issues intrusive to interdisciplinary approaches, we want to remind the readers that an understanding of human behaviors, and our chances to try and intervene in the dialogues that shape them, imply a strong commitment to confronting complexity which is exactly what entails any interdisciplinary approach. Clearly, this commitment might make our lives more difficult but refusing the opportunity means giving up on serious understandings that can pave the way for better educational designs.

We have organized the chapters in sections but the sections are inevitably interrelated and in some cases chapters could have been included in more than one section. The chapter "Desegregation and Resegregation: The Legacy of Brown versus Board of Education, 1954" by Gallagher offers the necessary background, often neglected, of the history of integrated education in United States of America that, due to academic compartmentalization, is not always accounted for in countries where integration is adopted as an educational solution toward peace. It sets the scene for subsequent critical commentaries on integrated education by other authors. The following two chapters deal with the need for system-wide responses to conflict in a variety of postconflict societies. Using the examples of Northern Ireland and Cyprus, Johnson, in "Moving from Piecemeal to Systemic Approaches to Peace Education in Divided Societies: Comparative Efforts in Northern Ireland and Cyprus," emphasizes the need for systemic rather than piecemeal approaches to peace education, while highlighting the challenges inherent in such initiatives. In the third chapter "Multiple Realities and the Role of Peace Education in Deep-rooted Conflicts: The Case of Cyprus," Hadjipavlou uses the Salomon model of peace education and the politics of poetry to call for a shared curriculum for reconciliation in Cyprus. In the fourth chapter "Reconciliation and Peace in Education in South Africa: The Constitutional Framework and Practical Manifestation in School Education," Bray and Joubert explore how the values of reconciliation and reconstruction enshrined in the South African constitution are becoming manifest in a new desegregated school education culture.

The second section explores issues around integrated schools in South Africa, Northern Ireland, and Israel from the perspectives of teachers and/or students. Using a variety of discourses from the debates around race and desegregation, Vandeyar and Esakov explore the experiences of students in desegregated schools in South Africa in the chapter "Color Coded: How Well Do Students of Different Race Groups Interact in South African Schools?" While the institutional culture of the schools appears resistant to change, the students themselves, whose identities are being transformed, are pushing for progress. In the sixth chapter "Challenges in Integrated Education in Northern Ireland," McGlynn critically examines the challenges facing the integrated education sector in Northern Ireland. Based on a series of interviews with school principals and other key actors and using a multitheoretical approach, she concludes that there is a need to clarify the meanings and objectives of integration, particularly with regard to the management of difference. Likewise, in "Developing Palestinian—Jewish Bilingual Integrated Education in Israel: Opportunities and Challenges for Peace Education in Conflict Societies," Bekerman investigates the challenges faced by an integrated education sector, this time a much smaller one and operating in the high-conflict situation of Israel. The potential benefits of long-term bilingual coeducation are indicated as are the difficulties faced by the participants in this enterprise. Bekerman critiques the theoretical perspectives that have underpinned these integrated education efforts and calls for their further development. In "Peace Education in a Bilingual and Biethnic School for Palestinians and Jews in Israel: Lessons and Challenges," Nasser and Abu-Nimer report on a case study of one of the bilingual, binational schools in Israel by exploring the perspectives of the Palestinian and

Jewish teachers and students. While the benefits of coeducation for promoting the understanding of the other ethnic group are apparent, as Bekerman also agrees, the bilingual objectives are challenged by hegemonic forces. In the final chapter of this section, "Is the Policy Sufficient? An Exploration of Integrated Education in Northern Ireland and Bilingual/Binational Education in Israel," Hughes and Donnelly use contact theory to directly compare the nature of the contact experience in the integrated schools in Northern Ireland and in the bilingual, binational schools in Israel. Their study suggests a greater willingness to engage in shared learning activities around the more controversial aspects of difference in the Israeli schools and a tendency to avoid this in the Northern Ireland schools.

The third section brings together three chapters exploring issues around curricula and pedagogy for sustained peace education. In "Education for Peace: The Pedagogy of Civilization," Danesh describes a groundbreaking Education for Peace program now being extended to schools in 65 cities in Bosnia Herzegovina with a view to reconciling Bosniak, Croat, and Serb populations. In the next chapter, "Learning to Do Integrated Education: "Visible" and "Invisible" Pedagogy in Northern Ireland's Integrated Schools," Moffat interrogates recent research on integrated schools in Northern Ireland using Bernstein's model of invisible and visible pedagogy and relates this to the professional development of teachers. The final chapter in this section, "From War to Peace: An Analysis of Peace Efforts in the Ife/Modakeke Community of Nigeria," by Aladejana, investigates the contribution of school and university curricula to the current state of peace amongst the Ife/Modakeke community of Nigeria, a country not often represented in the literature.

There then follow two chapters on the theme of adult peace education. In "Toward Sustainable Peace Education: Theoretical and Methodological Frameworks of a Program in South Africa," Houghton and John explore the theoretical and methodological frameworks of an adult peace program in South Africa, drawing on a communities of practice model and emphasizing the importance of follow up work. In "Learning and Unlearning on the Road to Peace: Adult Education and Community Relations in Northern Ireland," Nolan critiques different approaches taken by educators toward adult peace education in Northern Ireland, including those operating within Enlightenment and cultural diversity paradigms.

The final section of this book addresses an area flagged throughout the volume as one deserving urgent attention worldwide, namely the education and training of teachers for peace building. In the challenging context of Rwanda, where many teachers actively participated in genocide, Njoroge explores the reconstruction of the psyche of teachers that must precede their ability to respond to the task of social transformation. The theory and practice of teacher education at a Rwandan educational institution is described and the challenges therein are identified in "The Reconstruction of the Teacher's Psyche in Rwanda: The Theory and Practice of Peace Education at Kigali Institute of Education." Koshmanova and Holm in "Post-Soviet Reconstruction in Ukraine: Education for Social Cohesion" address another country often absent from the peace education literature and undergoing a critical period of social reconstruction, that is, Ukraine. In particular, the authors explore three ways in which teacher educators could promote acceptance of ethnically diverse groups among student teachers, including the adoption of the U.S. model of academic service learning. In the final chapter "Teacher Preparation for Peace Building in United States of America and Northern Ireland," Carter reviews the foundations of peace education and approaches to teacher education that have been identified as essential to peace building, with a particular comparison of policy and practice in Northern Ireland and southern United States of America. Recommendations for the advancement of peace education in teacher training are critically discussed.

In reaching the end of this short introduction, we want to share with the readers two of the many conclusions we have reached through our sustained dialogue. The first has to do with the need to become more attentive to the profound influence socio-historical-political contexts have on the options open to implement peace education. Although theoretically driven, peace education is a practical enterprise that needs continuously to consider what it is that the present contextual conditions afford it to do and say. Within each of the geographical contexts represented in our initiative (Nigeria, Israel, North Ireland, Bosnia Herzegovina, etc.) we have come to recognize contextual differences that make each of our own experiences limited in any attempt at translation. Second, the dialogue we sustained points to the urgent need to pay attention to the voices of educators involved in peace educational efforts who, from their personal and professional experience, bring a wealth of knowledge. An examination of this knowledge can greatly contribute to our understanding of the ways to experience and creatively deal with the dilemmas and challenges we hope to help change, with the theoretical perspectives that guide educational work and with the practicalities of the educational implementation and the aims and goals of the curriculum. These experiences/descriptions/accounts can be denied or taken seriously. Success in educational tasks is not dependent on the sharpness of theorizing but on the everyday activity of those we intend to train and/or educate. When taking these experiences/descriptions/accounts seriously, both educational planners and researchers will be able to use them as constructs in the development of broader and better training and educational designs instead of perceiving them as obstacles.

As previously stated, this volume represents only a first step in what we hope will evolve into future initiatives aimed at strengthening research and practice into the area of sustained peace education that is much in need of expanding its paradigmatic, theoretical, and practical horizons. We hope this volume will encourage others to contribute to such developments.

# Section I
# Systemic Issues

# Chapter One
## Desegregation and Resegregation: The Legacy of Brown versus Board of Education, 1954

*Tony Gallagher*

Although no society is ethnically homogeneous, in some the tensions associated with ethnic divisions are more prominent than in others. In such contexts a key factor concerns the organization of schools and the implications of this for ethnic relations. Broadly stated practice might be seen as lying on a continuum, with one end characterized by a plurality of schools and the other end by plural schools. A plurality of schools implies that different communities can organize their own schools and that the separate school systems reflects the plurality of the wider society. The notion of plural schools implies that each school seeks to embody the diversity that exists within society and that, preferably, a single system of common schools exists.

Of course life is never so simple. In many cases the existence of separate schools may reflect inequalities between communities, with differential access to resources or opportunities. In extreme circumstances, such as apartheid South Africa, separate schooling is part of an elaborate mechanism for domination. Otherwise it is minority communities themselves who opt for separate schools, mainly to maintain aspects of identity or culture, including language or religion. While the operation of separate school systems as a matter of choice is fully compatible with international human rights standards (Minority Rights Group, 1994), separation can have wider consequences for the whole society (Gallagher, 2004). This issue was addressed directly in Britain in the mid-1980s and it was concluded that separate schools would damage the interests of the wider society, while recognizing that a system of common schools would be successful only if the schools genuinely offered recognition of the plurality of identities and cultures that comprised modern British society (Swann Report, 1985). This sentiment is, of course, fundamental to the concept of *e pluribus unum* that underpins the social purpose of the U.S. public schools (Postman, 1996).

Many of the chapters in this book address issues related to this dilemma and, in particular, the problems and possibilities associated with schools that seek to provide plural contexts within which a diverse body of students and teachers can work together successfully. In all discussions on this issue a seminal experience is provided by the struggle to desegregate U.S. schools where, until 1954, many states obliged children of different "races" to attend separate schools—this is the focus of the present chapter.

It is half a century since the landmark decision of the U.S. Supreme Court that ruled legalized enforced segregation of young people into black and white schools to be contrary to the U.S. Constitution. The journey from that point has not, however, been a simple one. The decision did mark a sea change of enormous consequence in public policy in the United States of America and, eventually, it did have a significant impact on public opinion and practice. However, the most recent evidence suggests that, after several decades of progressive change, there has been a reversal of practice toward a pattern of resegregation in U.S. public schools. The purpose of this chapter is to examine the patterns of school practice in the years since 1954 and to explore some of the factors that have affected those patterns. Before doing this we briefly look at the background to the landmark decision.

> We hold these truths to be self-evident, that all men are created equal; that they are endowed by their Creator with certain inalienable rights; that among these, are life, liberty and the pursuit of happiness. (From the U.S. Declaration of Independence)

Most are familiar with the ringing rhetoric of the opening phrases of the 1776 Declaration of Independence and the revolutionary liberal claims that it espoused. Less obvious is the fact that the meaning attached to the idea of equality by the framers of the Declaration was highly constrained, by current standards, and reflected a view of the legal persona of "the citizen" that was largely limited to property-owning men. Thus, for example, the framers of the Declaration who held it to be self-evident that "all men are created equal" saw no contradiction between this assertion and section 2, article 1 of the U.S. Constitution that stated:

> Representatives and direct Taxes will be apportioned among the several States . . . according to their respective numbers which shall be determined by adding to the whole Number of free Persons, including those not bound to Service for a Term of Years, and . . . three fifths of other Persons. (From the U.S. Constitution, 1867)

Those deemed to be "three fifths of other persons" in this context were the slaves. Why three-fifths of a person? This was a compromise between those who wished not to count slaves at all, so that the tax bill would be reduced, and those who wished slaves to be counted to the fullest extent, in order to increase the number of representatives sent to congress from a state. From the perspective of the twenty-first century, however, we are left with a seemingly glaring contradiction in foundational documents, made more bewildering by the apparent blindness of contemporary observers (Tindall and Shi, 1999).

Legalized slavery in the United States of America did not survive the civil war, 1861–65. The Emancipation Proclamation freed the slaves and inaugurated the period known as Reconstruction when Northern radicals literally set about reconstructing the Southern states of the former confederacy in order to make them fit to rejoin the Union (Franklin, 1994). Reconstruction was marked by radical reform within the states, but also by new federal legislation in Washington DC and a series of constitutional amendments. For the present purposes the most important was the Fourteenth Amendment to the Constitution (1868) that included the following:

> No State shall make or enforce any law which shall abridge the privileges or immunities of citizens . . . nor . . . deprive any person of life, liberty, or property, . . . nor deny to any person . . . the equal protection of the law. (From the U.S. Constitution, 1868)

The last phrase in this statement, "the equal protection of the law," was to become the key focus of struggle for the nascent civil rights movement and continues today to provide the site of debate over the extent to which the state should intervene to affect the state of relations between communities.

The Reconstruction period did not last long. In a relatively short time a reaction to the radical agenda, and the Northern "carpet-baggers," set in across the South and was supported by a impatient federal government in Washington. Perhaps a key marker of this shift in mood and opinion was provided by the Supreme Court ruling in 1883 that the 1875 civil rights was *ultra vires* the constitution. The Civil Rights Act had deemed discrimination in the provision of goods and services on the basis of race to be illegal. The Supreme Court accepted the argument put to it that the Fourteenth Amendment to the constitution applied to government and could not be imposed on individuals who should be left free to decide their own relations with others. In the aftermath of this decision, which seemed to neuter the potential of the Fourteenth Amendment, a raft of cases provided tests of the extent to which segregation could be pursued, while remaining legal. The landmark decision emerged in the 1896 Supreme Court ruling, *Plessy v Ferguson*. This concerned a New Orleans case in which a black citizen refused to vacate a seat in a "white" railroad car to relocate in a designated "black" car. The court ruled that a requirement that citizens use separate facilities could be consistent with the constitution as long as the separate facilities were broadly equal (Baker, 1996; Polenberg, 1980).

This "separate but equal" condition set the conditions for public policy for the next half century. In the immediate aftermath of the decision it provided a green light for the passage of state legislation mandating separate facilities for blacks and whites, and this happened with particular alacrity in regard to the provision of schools. Thus, by the 1950s 17 of the 21 states that had participated in the confederacy and the district of Columbia required black and white children to attend separate schools.

The main representative organization of black Americans, the NAACP (National Association for the Advancement of Colored People), sought to mount legal challenges to the "separate but equal" ruling. In part this decision was based on the pragmatic view that the Dixiecrats, the bloc of Democrats elected in Southern states, could block any attempt to pass new civil rights legislation through congress and constrain significant action by the executive (even allowing for the possibility that there was presidential will for action in the first place). With the legislative and executive branches of government unlikely to provide a fruitful avenue for action, that left the judicial branch as the one most likely to lead to results. In addition, however, the NAACP had to devise a legal strategy that would not have the consequence of further strengthening the legal basis of segregation. This implied that they should seek to chip away at the conditions around the "separate but equal" ruling, before attempting a direct assault on the fundamental basis of the ruling (Kluger, 2004).

It was on this basis that they pursued a series of cases that focused on a rather narrow set of circumstances: they decided that they should focus initially on higher education, rather than compulsory education, on the grounds that it would be harder for states to meet the "separate but equal" conditions in meaningful ways; and they focused, where possible, on law schools, on the grounds that Supreme Court justices were more likely to recognize the value of entry to such institutions. The strategy was pursued over many years, but there are a number of landmark cases. In *Missouri ex rel. Gaines v Canada*, 1938, the Supreme Court agreed with the contention that a separate provision had to be within-state and that it was not sufficient for a condition of equality that the state agreed to fund a place for a black student in an out-of-state college. In *McLaurin v Oklahoma State Regents*, 1950, the plaintiffs successfully argued that a requirement that a black student use entirely separate facilities within a college effectively impaired the student's ability to complete the course and therefore also did not meet the legal condition for equality. And in *Sweatt v Painter*, 1950, the court ruled that the establishment of an entirely new in-state alternative college, with limited facilities and faculty, was insufficient to meet the conditions of equality (Raffel, 1998).

Having chipped away at the conditions for the Plessy ruling, the direct assault led to the *Brown v Board of Education* ruling that deemed that a requirement to use separate facilities was, by definition, unequal:

> To separate some children from others of similar age and qualifications solely because of their race generates feeling of inferiority... Separate facilities are inherently unequal.... Therefore we hold that the plaintiffs... are... deprived of the equal protection of the laws guaranteed by the Fourteenth Amendment. (From the ruling in *Brown v Board of Education*, 1954)

From this point onward it was no longer possible to maintain laws that required children to attend separate schools on the basis of their racial background. But this did not mean that action followed immediately: in a second ruling in 1955, Brown II, the court called for the establishment of plans to desegregate schools "with all deliberate speed," but in the absence of any targets or goals little planning or action was evident (Schlesinger, 1978). Indeed, in some situations where black citizens attempted to exercise their constitutional right to attend public schools, mobs of white citizens tried to prevent them. Most infamously in the case of Little Rock, Arkansas, the governor mobilized the National Guard to block entry to the High School by seven black children and was only forced to back down when President Eisenhower sent in federal troops.

However, despite the opposition of the white public and the deliberate lethargy of policymakers, change did eventually start to occur. And once change started, then it proceeded apace. The key turning point was the passage of the 1964 Civil Rights Act, and later the Voting Rights Act, through Congress. Now the three branches of government were in alignment on the issue and the focus of attention shifted from what might be done, to how it could be done more effectively. Once again the Supreme Court played a key role in enabling action. Thus, for example, the 1971 ruling in *Swann v Charlotte-Mecklenburg Board of Education* authorized the use of school busing as a direct means to achieve desegregated schools. In 1973, a similar ruling in *Keyes v School District Number 1*, Denver, authorized the use of busing in a Northern State where segregated schooling had existed on the basis of tradition rather than legal requirement (Raffel, 1998).

Gary Orfield and others at the Harvard Civil Rights Project have suggested that the 1964 Civil Rights Act, allied with progressive court rulings in the late 1960s and early 1970s greatly tightened the pressure on school districts actively to pursue measures to desegregate their schools. This, in turn, had the consequence that the system of virtual apartheid that had existed in the Southern states not only disappeared, but by the 1970s the South had the most integrated schools system of any part of the United States of America (Orfield and Eaton, 1996; Orfield and Miller, 1998; Orfield and Yun, 1999). Clotfelter (2004) used a variety of sources to illustrate empirically the extent of change across different regions of the United States of America. In 1960–61, the proportion of black students in 90–100 percent nonwhite schools in the South was 100 percent, but by 1976 this had dropped to 22 percent. By contrast, in the northeast the proportion actually rose over the same period from 40 to 51 percent. In the country as a whole the proportion fell from 64 percent in 1968 to 36 percent in 1976.

The era of progressive change proved, however, to be short-lived, much like the Reconstruction period after the civil war. The new turning point came with the 1974 Supreme Court ruling in *Milliken v Bradley*. This concerned a desegregation plan that required the busing of pupils between the predominantly black school districts in the city of Detroit and the predominantly white school districts in the suburbs surrounding the city. The case for the cross-district busing plan was that a desegregation plan confined to the city was meaningless as there was, for all practical purposes, no one to desegregate with. The case against the busing plan was that schools districts in which there was no evidence of intent to segregate could not be compelled to participate in solving a problem that existed in another school district. In a 5–4 ruling the Supreme Court agreed with the case against the busing plan and struck out cross-district busing plans, unless fault could be attributed. This decision highlighted the role of political influence on the Supreme Court: the liberal judgments from 1954 onward had emerged from Justices appointed largely by the Truman, Kennedy, and Johnston administrations, whereas the swing votes in the Milliken decision had come from justices appointed by Richard Nixon. With the presidencies of Reagan, Bush, and Bush yet to come the Supreme Court was to take on an even more conservative face in the closing years of the twentieth century.

We have seen how Clotfelter measured the change toward desegregation up to the 1970s. His data also show the pattern of resegregation that followed. We saw that in 1976 only 22 percent of black students in the South were in predominantly nonwhite schools, but by 2000 this figure had risen to 31 percent. In most other regions of the United States of America, according to Clotfelter, the level of segregation fell a little further up to 1980, but then started to rise: in the country as a whole in 1980 the proportion of black students in predominantly nonwhite schools had fallen to 33 percent, but by 2000 this had risen again to 37 percent. If the level of segregation had fallen so dramatically in the 1960s and 1970s, why then did the pattern reverse in the 1980s and 1990s? Clotfelter's (2004) analysis of the data suggests two main mechanisms, one operating mainly in the Northern states and the other operating mainly in the Southern states. In the Northern states he suggests that the Milliken, 1974, decision opened the way to white flight from the central cities to the predominantly white suburbs and their predominantly white schools (Jacobs, 1998). Further, there was an apparent guarantee that no cross-district busing strategy would be imposed. Thus, in the Northern states the main constraint on pressure for the continued desegregation of the schools came from the consequences of residential segregation and the legal loophole this provided for white families.

In the South, on the other hand, school districts were, by tradition, larger than districts in the North and so the option of white flight to the suburbs was not so readily available. Indeed, Clotfelter speculates that school districts may have been larger in the South precisely because legal segregation in the past had obviated the need for whites to press for residential segregation within district boundaries. In this context Clotfelter's analysis suggests that many white parents opted to send their children to private, and predominantly white, schools and that this acted as the main constraint on the continuation of desegregation.

Research from the Harvard Civil Rights Project points to a number of other factors that seemed to be important in the overall trends (Frankenberg and Lee, 2002; Frankenberg, Lee, and Orfield, 2003). The increasingly conservative Supreme Court made a series of rulings that reduced the pressure on school districts to engage in active desegregation measures and made it easier for districts to remove themselves from court oversight. Another factor arose from the changing demographics of U.S. society, in particular the rapid growth of the Latino population: this community had never enjoyed a period of desegregated schooling, but a consequence of court rulings in relation to black Americans meant that Latino students experienced increased levels of de facto segregation over time. For the Harvard Civil Rights Project, however, perhaps the most important factor lay in the absence of significant leadership in the promotion of integration as a priority social and educational goal. This lack of significant leadership is identified in all three branches of government: the last constructive act by congress on integration was, it is argued, in 1972; the Supreme Court began limiting the pressure for desegregation in 1974 with the Milliken decision, and from 1991 onward had engaged in the active dismantling of desegregation plans; and, apart from a brief hiatus during the Carter administration, they argue that the executive branch has carried out no substantial enforcement of desegregation since the Johnson administration.

One consequence of the 1974 Milliken ruling was a suggestion that the problems facing the predominantly black city schools might be alleviated by the injection of additional resources and funds. Even if this comes dangerously close to raising the possibility of a new version of "separate but equal" (perhaps best characterized as "separate but compensated"), it does raise the question as to whether the gains from integration were so substantial that policymakers, and the public, should worry when integration does not occur. Advocates of desegregation have examined evidence on this issue and have argued that the gains are both substantial and tangible. According to evidence collected by the Harvard Civil Rights Project three areas of student outcome are positively strengthened by integrated classrooms. These include evidence of enhanced learning by all students, higher educational and occupational aspirations particularly among minority students and evidence of positive social interaction among members of different racial and ethnic backgrounds, with the prospect of this interactive networking surviving beyond the school years (Kurleander and Yun, 2002a, 2002b).

A recent large-scale qualitative study by Amy Wells and colleagues examined the retrospective experience and longer-term consequence of desegregation from amongst students who were in the largest wave of busing and desegregation plans. The evidence provided by Wells et al. (2004) provides a rare insight into the human experience of large-scale social change. Her respondents described how desegregation of schools did not always lead to desegregated classes: they remembered the application of tracking that often led to within-school segregation as minority students were more likely to be assigned to lower tracks, while white students were more likely to be assigned to higher tracks. In addition, many recalled a reluctance on the part of the predominantly white teachers directly to deal with issues of race and color on the basis of an official policy of "color blindness." Despite this, however, Wells et al.'s respondents valued the extent to which they had experienced diversity in their schools and felt that it had prepared them better for living in a diverse society. Furthermore, they lamented the fact that, consistently, they now lived in more segregated environments and their children did not have the same opportunity they had to experience diversity.

A reflection of the balance sheet over the 50 years since the *Brown v Board of Education* decision suggests there are some positives, some possibilities and some problems. On the positive side there is a consistent pattern that most U.S. citizens support the idea of integration, even if the evidence of actual integration is more mixed. Nevertheless this is important as it suggests that there could be a political climate to support positive action to promote integration. A second positive emerged in a surprisingly liberal Supreme Court ruling in 2003 on affirmative action (Coleman and Palmer, 2003). There has been a long-running debate on the merits of affirmative action in the United States of America since the 1960s and the situation was not aided by mixed messages from the Supreme Court. By the early 2000s, the legal condition was that selective universities could use race as a criterion in an affirmative action plan if such a plan represented a compelling interest for the institution. There was a consensus that evidence of past discrimination provided grounds for a compelling interest, but there was disagreement on whether the goal of developing and maintaining a diverse student body also represented a compelling interest. This was tested in two cases, *Gratz v Bollinger* and *Grutter v Bollinger*, both involving white students who had applied to the University of Michigan. In its 2003 judgments the Supreme Court ruled in favor of the idea that the goal of diversity did represent a compelling interest and therefore that race could be used as part of an affirmative action admissions plan. However, the court also ruled against the use of a points-based system in which applicants received a fixed number of points if they were from a minority racial category. The court may have been influenced by an Amicus brief presented by the U.S. Army in favor of diversity as a compelling interest, which itself arose from the painful experience of the Vietnam War in which significant problems had emerged from a situation where largely white officers tried to lead black soldiers.

However, whatever the relative weight of factors that swung the justices' decision, the legal clarification in the Michigan cases does underpin the aspiration toward promoting diversity as a social good, even if it constrained the most obvious procedures that might be used to achieve that goal.

The decision in the Michigan cases opens up the possibility that diversity might come to be used as a marker of school quality and sit alongside academic standards as a key criterion in judging the worth of schools. Such a development may require a rediscovery of the social purpose of the U.S. public school (Postman, 1996), although that is now perhaps less beyond the bounds of possibility than it appeared to be just a few years ago. In an era when parental choice is becoming ever more important in school policy, it might also be possible to amend choice programs in order that they act to encourage diversity, rather than encourage segregation.

However, there was another aspect of the fiftieth anniversary of the Brown ruling that called into question the viability of a renewed effort toward desegregation. This emerged as a consequence of research among African American scholars who seemed to call into question the bona fides of white America (Walker, 2004, 2005). On the one hand some of this research reminded us that the main burden of change from separate schools to desegregation had been borne by black communities: when two schools merged, invariability it was the black school that closed; when teachers were made redundant due to economies of scale, invariability the rate was higher among black teachers as compared with white teachers; when leadership positions in the desegregated schools were allocated, almost invariably white teachers were advantaged in the opportunities available to them. On the other hand some of this research casts a critical and skeptical eye over the experience of the past decades: for some scholars the years of effort to promote integration were stymied by the reluctance of many white parents to engage positively with the process. At various times this has been described as the advocacy of states' rights, exit to private schools, flight to the suburbs, commitment to local schools, or opposition to externally imposed busing. For some African American scholars this all looks like white racism and a refusal to send white children to sit alongside black children in the classrooms of America. This makes some wonder whether the black community should continue to rely on the good faith of white America, or rely on their own resources.

At its fiftieth anniversary, then, the legacy of Brown versus Board of Education appears to be somewhat mixed. On the one hand, the 1954 decision held out the promise of a new and different America within which diversity would be experienced, and celebrated, on a daily basis. On the other hand, the period when it seemed possible that this hope might be fulfilled proved to be all-too brief as two decades of achievement have been followed by two decades of retrenchment. However, there is little doubt that the Brown decision was one of the factors that helped launch the Civil Rights movement and all the gains achieved by that campaign (Garrow, 1993) and there have been some fundamental changes in attitude in practice across the country: even the site of one of the most controversial episodes of the desegregation struggle, Little Rock High School in Arkansas, has now been dedicated as a national monument. So, for all the frustrations and problems of the present, the United States of America is a very different place now, in comparison with the legalized apartheid of the pre-Brown period.

Viewed on a wider conspectus the experience of Brown does not, unfortunately, provide a simple answer to the some of the dilemmas raised by ethnic diversity. Perhaps the key question in regard to diversity in education lies in the choice between separation or integration, but on this issue the experience of Brown in the United States of America provides only partial answers. Despite the legal cover provided by the 1896 Plessy decision, legally mandated segregation is an affront to democratic standards. If nothing else, the Brown decision provides an inspirational example of a struggle for democracy and inclusion. The logic of the decision was for a move toward greater integration, but in this respect the wider lessons from Brown are much less obvious. The evidence shows clearly that a level of integration can be achieved particularly when segregation is based on legislation, not least because laws can be changed and, in a lawful society, most people will adhere to the law, eventually. However, when segregation is based on custom and practice, it provides a more amorphous target that is harder to pin down and harder still to reverse.

And this highlights another frustrating aspect of the legacy of Brown. An undoubted gain has been the sea change in attitudes in the United States of America to the extent that, rhetorically, almost everyone supports the principle of integration. However, it is difficult in fact to translate this aspiration into practice, as too many of those with the power to choose opt for separate environments. Indeed whereas the fiftieth anniversary of Brown was widely acknowledged and celebrated across the United States of America, it was and is striking just how few people still remain as unremitting advocates of its core integrationist message. The U.S. experience before Brown does point to the conclusion that forced segregation is ultimately self-defeating, not least because it is predicated on an unsustainable vision of the essential separateness of human groups. Furthermore, there is some evidence from the period of desegregation in the United States of America that suggests that integration can lead to tangible gains. However, the experience does not suggest that integration is the unproblematic answer to diversity: integration was hard to achieve, harder still to sustain and harder again to maintain. The current state of education in the United States of America is one in which the celebration of diversity stands as a common goal, but the practice of (benign?) separation stands as a lived reality for many. So, if the key question to ask of Brown is: does it prove that integration is the solution, then, unfortunately, the best we can answer at the moment is, "maybe."

The legacy of Brown perhaps also points to another conclusion. The success of the drive towards desegregation in the 1960s, particularly in the Southern States where legalized segregation had applied, indicates the value of forcing behavior in socially progressive directions through the use of legislation. However, focusing on behavior is predicated on the assumption that behavioral change will be followed by cognitive and affective change—in other words, if you force people to do good things then, in time, they will come to believe and feel that they should be doing only good things. In this way, it is hoped, change will be sustained. However, the pattern of resegregation that is occurring in the United States of America raises some fundamental problems with this expectation. It suggests that changing behavior alone may not provide a sufficient basis for sustained change, but that additional efforts on the cognitive and affective dimensions need also to be pursued. There have been other contexts in which educators have worked through schools to promote shared educational contexts for young people from diverse ethnic or religious backgrounds. The experience of some of the these other contexts are considered in the present volume, but it is perhaps the experience of such efforts in Israel and Northern Ireland that are given particular focus: in both settings a minority of educators have advanced the idea that shared schooling provide an important basis for promoting shared living, but their efforts have been characterized by a commitment to the idea that shared schools should be more than places where different students come together. Rather they should be qualitatively different contexts within which the curriculum and ethos of the school deals directly with the issue of diversity. They are contexts, in other words, where there is a commitment to addressing the cognitive and affective every bit as much as the behavioral. The following chapters in this volume consider how they have pursued this and how successful they have been.

## References

Baker, L. (1996). *The second battle of New Orleans: The hundred year struggle to integrate the schools.* New York: Harper Collins.

Clotfelter, C.T. (2004). *After Brown: The rise and retreat of school desegregation.* Princeton and Oxford: Princeton University Press.

Coleman, A. and Palmer, S. (2003). The U.S. Supreme Court decisions in Gratz v. Bollinger and Grutter v. Bollinger: Case analysis and lessons learned regarding the use of race by colleges and universities. Retrieved July 1, 2003 from www.nixonpeabody.com

Frankenberg, W. and Lee, C. (2002). *Race in American public schools: Rapidly resegregating school districts.* Cambridge, MA: Harvard Civil Rights Project.

Frankenberg, W., Lee, C., and Orfield, G. (2003). *A multiracial society with segregated schools: Are we losing the dream?* Cambridge, MA: Harvard Civil Rights Project.

Franklin, J.H. (1994). *Reconstruction after the Civil War.* Chicago: University of Chicago Press.

Gallagher, T. (2004). *Education in divided societies.* London: Palgrave/Macmillan.

Garrow, D.J. (1993). *Bearing the cross: Martin Luther King Jr. and the southern Christian leadership conference.* London: Vintage.

Jacobs, G.S. (1998). *Getting round Brown: Desegregation, development and the Columbus Public Schools.* Ohio: Ohio State University Press.

Kluger, R. (2004). *Simple justice: The history of Brown v. Board of Education and Black America's struggle for equality.* New York: Vintage.

Kurleander, M. and Yun, J.T. (2002a). *The impact of racial and ethnic diversity on educational outcomes: Lynn, MA school district.* Cambridge, MA: Harvard Civil Rights Project.

———. (2002b). *The impact of racial and ethnic diversity on educational outcomes: Cambridge, MA school district.* Cambridge, MA: Harvard Civil Rights Project.

Minority Rights Group. (1994). *Education rights and minorities.* London: Minority Rights Group.

Orfield, G. and Eaton, S. (1996). *Dismantling desegregation: The quiet reversal of Brown versus Board of Education.* New York: New Press.

Orfield, G. and Miller, E. (1998). *Chilling admissions: The affirmative action crisis and the search for alternatives.* Cambridge, MA: Harvard Education Publishing Group.

Orfield, G. and Yun, J.T. (1999). *Resegregation in American schools.* Cambridge, MA: Harvard Civil Rights Project.

Polenberg, R. (1980). *One nation divisible: Class, race and ethnicity in the United States since 1938.* Harmondsworth: Penguin.

Postman, N. (1996). *The end of education.* New York: Vintage.

Raffel, J.A. (1998). *Historical dictionary of school segregation and desegregation: The American experience.* Westport: Greenwood Press.
Schlesinger, A.M. (1978). *Robert Kennedy and his times.* New York: Ballantine Books.
Swann Report. (1985). *Education for all.* London: HMSO.
Tindall, G.B. and Shi, D.E. (1999). *America: A narrative history.* New York: W.W. Norton & Company, Inc.
Walker, V.S. (2004). *The implications of the brown decision for educational reform.* Paper presented at the annual AERA conference, San Diego, CA.
———. (2005). Organized resistance and black educators' quest for school equality, 1878–1938. *Teachers College Record,* 107 (3): 355–388.
Wells, A.S., Holme, J.J., Revilla, A.T., and Atanda, A.K. (2004). *How desegregation changed us: The effects of racially mixed schools on students and society. A study of desegregated high schools and their class of 1980 graduates.* California: Teachers College, Columbia University/ University of California at Los Angeles.

# Chapter Two
# Moving from Piecemeal to Systemic Approaches to Peace Education in Divided Societies: Comparative Efforts in Northern Ireland and Cyprus

*Laurie Shepherd Johnson*

In a world increasingly besieged by geopolitical and interethnic conflict, the notion of societal peace has become ever more elusive. Despite these conditions, a growing number of postconflict states are turning from violence to political diplomacy as a means of moving beyond the enmity that has divided them. Nonetheless, relying on politics as the primary channel toward peace in these regions remains less than satisfactory. In divided societies like Northern Ireland and Cyprus, where deeply entrenched distrust of "the other" has impeded political progress toward peace settlement, and governmental power-sharing attempts have been blocked and efforts to promote interdependence across the divide are often censured, it is clear that political diplomacy alone will not mend the walls of division.

In divided states where people have experienced violent conflict, both sides hold on to their perception of the other as the enemy by tirelessly venerating their own "chosen traumas" and "chosen glories" (Volkan, 1998). Reconciliation becomes near impossible when this psychological stance persists. When people continue to harbor feelings of injustice in relation to "the other," as in postconflict zones where acts of previous violence have destroyed lives, property, and dignity, it is very difficult to negotiate a peaceful coexistence (Zuzovski, 1997). If sustainable peace is to be built in these regions, a dramatic change in the collective worldview is needed, a reframed understanding of "the other" must be developed, a sense of restorative justice achieved, and the insular systems that serve to propagate division must be transformed. In the end, a sustainable culture of peace in these regions will only be achieved through broadly based initiatives that seek to promote social cohesion in developmental and systemic ways.

Once the goal of rapprochement becomes part of the societal discourse and diplomatic means to settlement are underway, education stands as the natural vehicle through which social cohesion can be pursued over the long haul. Education, as a primary conduit for the transmission of knowledge, culture, and values, serves as the medium through which society collectively comes to define itself. The problem often is that school systems in divided states serve to strengthen, if not promote, the divisions. It has been argued, for example, that the educational systems in both Northern Ireland and Cyprus, segregated along religious, cultural, and geopolitical lines, have served to reinforce the societal conflicts in these respective states. In these regions, the vast majority of students study in segregated single-identity environments where they come into contact only with "their own," where their classmates, teachers, and role models come from the same sociocultural background, where ethnocentric versions of history are promulgated and one's own cultural heritage and identity are considered preeminent. In these systems, hostilities toward "the other" can proliferate, if not as the result of deliberate acts, then as the result of benign acceptance of longstanding prejudices. Characteristically, single-identity schools do not provide meaningful opportunities for students to achieve mutual understanding, respect, and interdependent cooperation across the divide (Church, Johnson, and Visser, 2004).

Without the opportunity for young people to learn how to live cooperatively with "the other," how does a divided society move toward reconciliation? Through what collective means can efforts be taken to promote knowledge about "the other," foster an appreciation of the values of mutual respect, equity and tolerance, and build the skills and behaviors needed to live and work interdependently in an increasingly diverse world? Increasingly, educationalists around the world are recognizing that these outcomes can best be achieved by pursuing a systemic approach toward educating for a culture of peace.

Peace education, however, can succeed in promoting social cohesion only when it is systemically integrated and politically contextualized. Beyond adding rights-based content and participatory processes aimed at individual learners, peace education must also engage the broader society and its systems. Stand-alone curricula and learning activities that are not part of an integrated system of the whole are bound to fail at educating for sustained peace, most especially in regions where deep-seated fears and mistrust have, over the decades, infiltrated collective ways of being. While peace education seeks to promote on the part of the school learner the transmission of knowledge, attitudes, and skills needed to attain and maintain peace, justice, security, and environmental sustainability, it must also be about developing societal capacities in the broader community.

Systemic approaches to peace education must include concerted engagement at multiple levels of government, education ministry, political party systems, labor/teacher unions, commercial enterprise, school and university, and family and community. System-wide strategies, policies, and structures need to be established, from the individual school level up through the national ministry. Curriculum development and implementation at the highest levels, including revision of curricular materials to better assure inclusion and parity across cultural heritage, national identity, gender, class, and politics needs to be undertaken. Similarly, teacher training and pedagogy need to be reconceptualized and restructured at the national level in ways that will further ensure that educators will be equipped to promote the knowledge, dispositions, and skills needed for a culture of peace.

## Education in the Context of Northern Ireland

Northern Ireland (NI) is a divided society with an entrenched history of sectarian conflict between its Protestant/Unionist and Catholic/Nationalist communities. This division is played out and reinforced in every aspect of Northern Ireland society, from housing to employment to recreation, and most especially illustrated in its segregated educational system. Children attend separate schools by religious and cultural tradition; as such, longstanding sectarian attitudes continue to be tacitly reinforced in the very institutions that are intended to prepare young people for success in the future. About 95 percent of children in Northern Ireland are educated in schools that would be considered "single-identity." The remaining 5 percent attend independent schools or integrated schools, the latter of which now represent a growing presence in the educational system (see the discussion shortly).

## Education in the Context of Cyprus

The island of Cyprus has been a divided society ever since violent hostilities between its Greek and Turkish communities in 1964 led to the assignment of United Nations peacekeepers on the island and then to formal division in July, 1974, when Turkey invaded the northern part of the island in response to a pro-Greek coup attempt to unify Cyprus with Greece. Over time, the division on the island has become complete: politically, geographically, culturally, and psychologically. The bifurcation is made all the more explicit by the "Green Line" that cuts across the center of the capital city of Nicosia and extends 180m across the island, dividing Greek Cypriots (GC) who live in the southern part from the Turkish Cypriots (TC) who live in the northern third of the island. As a result of these physical and psychological barriers, the vast majority of Cypriots have had no meaningful contact with members of "the other" community during this time (Constantinou and Papadakis, 2001).

The education system has been criticized as playing a role in sustaining and perpetuating the conflict in Cyprus where division is reinforced not only through the texts students are given to study but by the ethnocentric tenets that are espoused in the classroom (Hadjipavlou-Trigeorgis, 2000). Cypriot children are educated in separate school systems according to the geopolitical and ethnic lines of division. In 2002–03, approximately 123,000 GC students (Ministry of Education and Culture, 2003) were attending schools in the southern two-thirds of the island where the primary language and ethos is Greek and approximately 27,000 TC students attend schools in the north (personal communication on January 16, 2004 with TC educational authorities in North Nicosia) where the primary language and ethos is Turkish. Adding to the layers of division in this society, a growing influx of immigrant and refugee children has been occurring in GC schools over the past decade with a corresponding increase in localized reports of ethnic hostility noted by these schools (Trimikliniotis, 2003).

## Systemic Peace Education Approaches in Divided Societies

Postconflict societies struggle against unique challenges in relation to implementing initiatives aimed at rapprochement; cross-community resistance is great and obstacles abound at every level. The divided societies of Northern Ireland and Cyprus serve as good illustrations of this. Comparatively, the educational system of Northern Ireland has advanced more in its efforts to conceptualize and implement systemic peace education. While not without resistance, structural, curricular, and policy initiatives in Northern Ireland are taking place that support peace education goals on a system-wide basis. In Cyprus, where political settlement issues continue to take center stage, there has been demonstrably less readiness to pursue anything that would reflect system-wide educational efforts aimed toward social cohesion. Recent accession to the EU and the need to abide by European Council standards, however, are now serving as incentive to advance implementation of such initiatives.

In making comparisons between Northern Ireland and Cyprus, it is of course important to remember that, as internally divided societies, each respectively comprises disparate constituencies and, as such, should not be considered as nation-states with one national identity in traditional terms. Given the dichotomous contexts of these states, unilateral comparisons become complicated. Keeping this in mind, it is still useful to look at how the educational systems of Cyprus and Northern Ireland address matters of organizational structure, curriculum, and teacher training in the effort to promote peace in their societies.

## Organizational and Structural Initiatives

### Integrated Education

A small but vocal minority in Northern Ireland has been arguing for 30 years to establish integrated schools as a means of providing an alternative to the segregated system of education in this divided society. In 1981, the first integrated school in Northern Ireland was established as the result of a parent-driven, grassroots initiative to provide young people the opportunity to be educated together within an institutional ethos that promotes mutual understanding, respect, and interdependence.

Following slow but steady growth over the past 25 years, there are now 58 integrated schools in Northern Ireland (Retrieved May 21, 2006 from www.nicie.org) that enroll about 5 percent of the school-aged children in this province (NICIE, 2004). Over these years, the option of integrated education has become increasingly attractive to a larger percent of the population; in 2004, an estimated 700 families had to be "turned away" from integrated schools that were at full capacity (personal correspondence with NICIE chief executive officer, October 4, 2004).

The growth of Integrated Education (IE) has been hard won in the face of considerable resistance, not the least of which has come from religious quarters (Gallagher, Osborne, and Cormack, 1993; Leichty and Clegg, 2001; McDonald, 2002; O'Connor, 2002). Strong opposition has come from the claim of the church that educational integration seeks to curtail expression of religion and threaten "cultural identity." Both Catholic and Protestant hard-liners are known to openly denigrate families who send their children to an integrated school as "traitors" to their respective community (Johnson, 2002). Equally unsettling are those reports from integrated school managers (in schools located in flashpoint areas) who have been victimized by paramilitary group members striking out against integration with intimidation tactics aimed at preserving sociopolitical hegemony in their respective communities (private communication with a senior manager of integrated primary school, March, 2001).

The conciliatory impact that IE has had in Northern Ireland is perhaps best demonstrated by the fact that, in a recent social attitudes survey, three quarters of parents indicated they would opt for integrated education if more integrated schools were available in their community (Gallagher and Smith, 2002). Overall, research that has sought to assess the efficacy of IE in Northern Ireland has generally demonstrated affirmative findings (Johnson, 2002; McClenahan et al., 1996; McGlynn et al., 2004; Montgomery et al., 2003). Johnson assessed the practice of integrated education from the perspective of teachers across nine different integrated schools at both the primary and secondary levels. The preponderance of input in this study substantiated integrated education as successful in fostering learning environments in which individuals, students, and staff, from different backgrounds come together, typically for the first time in their lives, and learn to accept each other and interact cooperatively in a safe and reliable community. Most of the teachers in this study attributed success to the "whole-school climate of support and encouragement" that is strategically provided for students in integrated schools. These teachers indicated that the commitment to inclusion and to fostering self-esteem and respect on the part of all members of the school community served as the foundation of their daily practice. This study could not substantiate however whether these cooperative student behaviors extended beyond the campus at the end of the day.

In 2004, McGlynn and colleagues summarized the relevant research on integrated education in NI and found overall that IE positively impacts identity, out-group attitudes, forgiveness, and reconciliation. With all of its growing pains, the integrated education enterprise in Northern Ireland stands as an outstanding example of how societal change can be systematically fostered in regions plagued by protracted conflict and division.

In the case of Cyprus, instituting integrated schooling does not appear a viable option at this time. Due to the very real geopolitical boundaries that divide Cyprus and separate its GC and TC communities, establishing integrated schools as a means of systemic peace education is not practical on this island. Until the border is fully opened and people are free to move to and from north to south, the notion of teaching all children together is seemingly light years away. This being said, I argue that it is feasible to consider capitalizing on the opportunity of having the few exceptional schools where GC and TC students do currently study together in Cyprus (and where English is the common language) move further toward systemic commitment to integration, as time and politics progress. Among the schools in Cyprus that have integrated enrolments, the most noteworthy is the English School located in the capital of Nicosia where the school's original charter to educate both TC and GC students together has been reinstated after 30 years of hiatus generated by the events of 1974. Over the past two years, this school has been valiantly working toward building a culturally responsive school ethos in which all students are educated in an environment that promotes respect, interdependence, and mutual understanding.

My research at this school has substantiated its value as a model for systemic integration in an otherwise completely divided society; its efforts are now being incorporated into workshops on "best practices" offered to educators throughout the island.

Other than these examples of school integration, continued efforts are being made by committed teachers, NGOs and community activists to promote bicommunal contact schemes where TC and GC students are brought together on planned occasions and given the opportunity to interact with each other, similar to the earliest cross-community contact goals of Northern Ireland's "Education for Mutual Understanding" initiative (see discussion later).

## Curricular Initiatives

Curricular development initiatives are foundational to implementing systemic peace education in a divided society. Rather than relying on longstanding curricula that may present myopic perspectives, curriculum revision efforts that aim to broaden the learners' understanding of their society, its people, and its interrelationship to the world are needed at the national level. This includes undertaking system-wide efforts to assess all current texts and teaching materials that are being used in schools for indications of relevant revision. Equally important is working to develop educational sources and experiences that aim to provide students exposure to the knowledge, attitudes, and skills needed for living in a culture of peace in areas such as conflict resolution, collaborative problem-solving, and cross-cultural communications, for system-wide change to be instituted, these efforts need to include mutual involvement across the separate communities.

### Cross-Curricular Statutes

In an early educational reform effort to address the divisions within its society, Northern Ireland required, as part of the Educational Reform Order of 1989, that all of its schools institute the "Education for Mutual Understanding" (EMU) cross-curricular theme aimed at promoting cross-community learning and contact activities for students. While noteworthy as a nascent effort to promote cross-community contact among school children in sectarian Northern Ireland, research has indicated that implementation of EMU was unreliable, partly due to the lack of implementation standards that accompanied it (Smith and Robinson, 1996). While less than fully effective in carrying out its intended goals, the EMU cross-curricular initiative in Northern Ireland nonetheless produced certain gains in promoting the role of the educational system in fostering cross-community relations and in enhancing awareness of the need for formal initiatives to educate for peace and reconciliation in this society. The lesson learned from the EMU experience is that when a centrally directed curricular initiative is not secured by formal, national requirements supported by system-wide implementation standards, training and funding, even the best of intentions will not be sufficient to sustain it in a divided society. However, when looking comparatively at other divided nations like Cyprus where such effort is yet to be attempted, it is difficult to argue against the transitional value that such system-wide statutes can have as a means of breaking the ice and moving, if ever so slowly and imperfectly, in the direction toward cross-community contact.

*Citizenship Education (CE)*
Education serves as a conduit by which a society transfers its history and shapes its future. While education has been influenced by ethnocultural preservationist goals throughout history, the growing model in contemporary Western Europe is to utilize the education system as a medium for promoting students' understanding of their broader rights and responsibilities as citizens in an increasingly diverse and interconnected world. There is an increased understanding of the need to develop and foster knowledge, attitudes, and skills that will enable young people to play their part in building a more equitable society and interdependent world. As part of this, the Council of Europe officially declared 2005 as the "European Year of Citizenship through Education" and thus promoted CE as an especially promising vehicle for peace education in divided societies.

Over the past five years, significant energy has been devoted to the development and implementation of citizenship education in the mandated curriculum of Northern Ireland. Not surprisingly, the system-wide introduction of a citizenship curriculum has generated much debate. In a divided society such as Northern Ireland, any educational initiative that would seek to promote an expanded worldview beyond that which has been framed by its longstanding conflict narratives would expectedly encounter considerable resistance. Smith (2003) specifically notes the difficulties inherent in considering the concept of citizenship in a society where there are different loyalties that give rise to conflict; realistically it would seem that neither the British nor the Irish national identity provides the basis for a citizenship model that would be acceptable to all in Northern Ireland.

Beyond the debates, a significant corps of educationalists, practitioners, community NGOs, government officials, and policymakers have worked to develop a viable curricular model that would address the longstanding need to better educate Northern Irish students for responsible local and global citizenship. As part of the model's development, a large number of schools (representing the controlled, maintained, and integrated sectors) participated in the Social, Civic, and Political Education pilot project (SCPE) that was carried out by the University of Ulster and the Council for the Curriculum, Examinations and Assessment (CCEA). This project produced the conceptual framework for the Local and Global Citizenship program curriculum that will become a statutory provision in the national curriculum in 2006.

As it stands now, citizenship will be introduced as part of a "Learning for Life and Work" curriculum (compulsory at postprimary level) and will look at concepts centered around four themes: diversity and inclusion, human rights and social responsibilities, equality, and social justice, and democracy and active participation (Retrieved October 17, 2004 from http://www.deni.gov.uk/de_news/press_releases/sep_04/07.09.04.html). Each of the four themes are explored within local, national, European, and global contexts (Douglas, 2003). It is anticipated that this new approach to citizenship education might be seen by some teachers to pose a threat to the integrity of individual subjects. "In many respects it challenges the more traditional approach to teaching and learning as it seeks to engage pupils in an enquiry-based approach" (Douglas, 2003, p. 15). These concerns point to the great need for teacher training as part of implementation efforts (see discussion later).

The latest proposals for a revised Northern Ireland Curriculum are presented on the website of the CCEA at www.ccea.org.uk. Those familiar with the proposed curriculum, suggest that while there is a rights-based element to the curriculum it falls short of being considered human rights education. Despite encouragement from the Human Rights Commission of Northern Ireland (NIHRC), the incorporation of rights-based pedagogy into the national curriculum has remained contentious. Many in the unionist community associate the call for human rights with the nationalist community's long-term civil rights efforts, and thus view rights-based education as acquiescing to nationalist ideology.

In the Republic of Cyprus, educational reform efforts aimed at bringing the educational system into compliance with the precepts and standards of the Council of Europe have paralleled its accession to the EU. The Ministry of Education and Culture in the Republic of Cyprus states that it "sees among its priorities the European dimension of education and is taking all measures to ensure that Cyprus has an active role in education issues within the European Union." Corresponding with EU accession, "proposals regarding citizenship education, human rights education, and peace education have been made and are now under review by various stakeholders" (personal correspondence with the Office of the Acting Director of Primary Education, October 6, 2004).

Historically, civic education in Cyprus has been infused into the secondary education subject areas of History, Civics, Greek Literature, and Philosophy, and has for the most part remained traditional in its content coverage and pedagogical methods. A few years ago, an expert committee was appointed by the government to make a comprehensive educational reform proposal that has led to recommendations that have been posted on the Ministry of Education and Culture's website (www.moec.gov.cy). However, these recommendations are presented with no obligations or mandate for implementation in schools. In its 2003–04 annual report on General Secondary Education, the Ministry of Education and Culture denoted no reference to citizenship education in the curriculum but identified among its educational goals for the 2004–05 year "developing an active democratic citizen" and promoting "bicommunal cooperation" as part of its postaccession efforts to emphasize Cyprus' role in an United Europe (Retrieved July 22, 2005 from www.moec.gov.cy). To date, there is no system-wide curricular implementation of citizenship education in GC schools.

*Textbook Revision*
The use of a textbook in a particular subject area implies that an agreed knowledge base has been determined as that which students need to know in the given subject matter. On a certain level, textbooks serve an academic gatekeeping role in society. As such, textbooks represent fertile ground for promulgating the national narrative in divided societies. In particular, history textbooks have been instrumental in transmitting the dominant culture worldview to the current generation of school learners. When texts present information from only the dominant perspective, students stand to suffer from exposure to a biased construction of knowledge. Educating for a culture of peace involves developing textbooks, as primary instruments for learning, that speak beyond the dominant narrative. In 1995, UNESCO declared that textbooks should be cleared of

negative stereotypes and when presenting issues should promote a sense of "otherness" while offering multiple viewpoints based on scientific facts, not national or cultural background (Pingle, 1999). Textbook revision projects are moving to front stage in several postconflict societies as part of educational reform efforts to better ensure more balanced and inclusive accounts in the education of youth regarding subject narratives, most commonly in history texts. Educational curricula in postconflict states need to move away from the traditional monocultural perspective and move toward "multiperspectivity" that recognizes that a multiple range of interpretations may exist in history, especially over large and controversial events, and that because different groups of people see historical events differently, multiple perspectives can be legitimate. It has been argued that developing multiperspectivity in history curricula is the only way in which controversial issues can be discussed in a peaceful manner (Retrieved October 4, 2004 from www. gei.de/english/projekte/southeast/report_ann_low-beer.shtml). Member states of the EU are increasingly responding to the Council of Europe's call for multiperspectivity in history teaching.

In Northern Ireland, there have been efforts toward promoting a multiperspective orientation in history textbooks. However, a systemic initiative to assess all curricular materials for elements of societal bias is yet to take place.

In Cyprus, there were efforts initiated in the period leading up to EU accession to review the history curricula and texts; however, in the shadow of the 2004 defeat of the Annan Plan referendum and the corresponding resurgence of ethnocentric and nationalistic sentiments on the ground, these efforts have been slowed. While the Republic of Cyprus has signed on to the Council of Europe's call for history textbook review, official work toward this end has largely been postponed by the Ministry of Education and Culture. There are a number of educationalists, however, who are persisting with goals to develop revised texts and teaching guides of Cyprus' history, some of whom are seeking the collaborative input of the Council of Europe regarding best practices in teaching history in pluralistic societies.

### Teacher Training Initiatives

Classroom teachers in divided societies commonly come from single-identity backgrounds and have been trained in single-identity institutions where they have had little exposure to pluralistic worldviews (Johnson, 2002). As such, there is a great need to train these teachers in ways that will expand their knowledge, attitudes, and skills in critical areas that will equip them to educate children for a culture of peace. Method and content training in antidiscrimination, conflict resolution, and social justice need to be part of this, including learning pedagogical methods in group and participatory strategies that may not have been part of traditional training models. Being trained in cooperative learning strategies and "academic controversy" debate practices, for example, will enable teachers not only to enhance student learning as has been borne out in research, but will equip them to help build skills essential to peace building such as conflict resolution and collaborative problem-solving (Johnson and Johnson, 1994). In these ways, teachers can become "key agents" in the social reconstruction process of "achieving equality and justice in society at large" (Furlong, 1992, p. 167).

In Northern Ireland, there has been less than sufficient commitment at the state level to instituting the necessary frameworks for teacher training in support of peace education initiatives. This insufficiency was borne out, for example, in the fact that preservice teachers received only minimal exposure to the principles and implementation goals of the EMU (Johnson, 2002). There has been a system-wide lack of training in interculturalism, conflict resolution and social justice issues in NI's teacher preparation curriculum overall. Even in the integrated schools, which by their very nature are committed to pluralism, teachers have complained of the lack of training in these areas (Johnson, 2002). This being said, the Department of Education of Northern Ireland has recently earmarked funding for teacher training as part of its specific initiative to implement citizenship education in the national curriculum. As such, postprimary schools are now sending selected teachers to participate in a training program orchestrated through the five regional Education and Library Boards (ELB's) to prepare for the new curriculum provision. Furthermore, the state provided funding to establish the post of citizenship officer in each of the ELBs to facilitate the process (Douglas, 2003). The intention was to make the citizenship provision statutory after all schools participated in an in-service training program funded by the Department for Education.

Since accession to the EU, Cypriot educators have been trying to determine how best to align with a European future while remaining loyal to cultural heritage and national identity. Accordingly, the notion of educating teachers "for a new world" is slowly entering the professional education discourse. Teacher training in the areas of multiculturalism, tolerance, human rights, conflict resolution, and social justice, however, has not been a systemically recognized goal. In response to historical divisions, Phtiaka (2002) argued that teacher education in Cyprus "needs to concentrate on the future" by educating teachers to cultivate in themselves and pass on to their pupils a deep knowledge of history, a broad social conscience, tolerance toward all kinds of differences, a good knowledge of the TC people and culture as well as knowledge of other peoples and their cultures, and a deep commitment to social justice (pp. 362–363).

Clearly, the onus cannot rest on the teachers' shoulders. The teacher education and training institutions need to take on the development of standards and implementation of curricula that will provide preservice and in-service education in the areas pertinent to peace education. Over the past two years, there have been delimited efforts, in both the TC and the GC educational systems, to train teachers in intercultural and antidiscrimination pedagogy. Systemic commitment toward implementing curricular statutes in these areas, however, appears far off, and many would argue, will remain untenable until a political settlement is attained on the island.

## Labor Union Initiatives

Teachers unions are growing in power and influence in many European states (Robertson 2000; Tomlinson and Pashiardis, 2002). Their latent role in achieving systemic peace education is critical in that they are increasingly able to define chief items in negotiating the role and practice of teachers. However, the range of issues and scope of influence teachers unions hold sway on varies by region and political dynamics. In Cyprus, the TC teachers' union in the north has been noted for increasingly gaining power in the political sphere and for being successful in influencing the state agenda in favor of settlement with the GC community in the south (Yusuf, 2003). On the other hand, in the south of Cyprus, while the GC teachers' unions are increasingly being recognized in policymaking processes, their range of influence still primarily remains in traditional school-based matters such as those defined by educational reform objectives including negotiating the role and appraisal of teachers (Tomlinson and Pashiardis, 2002). In NI, the teachers' union at the primary level has been instrumental in promoting cross-border initiatives, in partnership with the teachers' unions in the south of Ireland (ROI) and with Amnesty International, to promote human rights education in primary schools throughout Ireland (Retrieved July 22, 2005 from http://www.amnesty. org.uk/ni/liftoff/ about.shtml); however, at the secondary level the teachers' unions appear more cautionary in taking on social cohesion issues in their policy pursuits. While the relative range of influence in terms of policymaking is variable, the engagement of teachers unions (in Western nations especially) is critical to secure as part of any systemic peace education initiative.

## Conclusion

In postconflict societies like Northern Ireland and Cyprus, educational reform efforts that are aimed at bridging the societal divide typically meet with great resistance from traditionalists who want to retain the status quo in the name of preserving cultural heritage and national identity.

In many ways, Northern Ireland has come further along in this regard than Cyprus within the same 30-year period. Bolstered by formal diplomatic efforts toward peace agreement in the society at large, the educational system of Northern Ireland has been able to pursue many incremental peace education efforts that have helped to move its society along thus far. Perhaps some of this is due to the fact that reconciliation and restorative justice efforts have had a longer history in Northern Ireland, as illustrated by the public inquiries that have been established such as the Bloody Sunday Tribunal running over the past six years (http://cain.ulst.ac.uk/events/bsunday/bs.htm). Cyprus, on the other hand, has had only sporadic and isolated opportunities to explore the conflict and its human implications from the perspective of both the communities. It is hoped that, with the partial lifting of the border in April 2003, opportunities for reconciliation are expanding. Since then, thousands have crossed the border daily and emotional meetings between former friends and neighbors have helped overcome some of the prejudices (Retrieved December 17, 2004 from http://news.bbc.co.uk/1/hi/world/europe/2839603.stm). At the same time, old wounds have been opened for many. As an island onto itself, Cyprus is more insulated than Northern Ireland which has benefited from the pushes and pulls of the European political landscape (and longer EU affiliation). Additionally, in Cyprus the cultural differences between GC and TC communities (e.g., language, religion, and tradition) are arguably greater than those experienced between Catholics and Protestants in the Northern Irish setting. All of these factors differentially impact the viability of peace education in these regions.

In a postconflict society, pursuit of peace education aimed at promoting social cohesion requires sensitivity to the nuances that define the given conflict, politics, and cultural groups in that unique context; there is no "one size fits all" model for learning how to live together in broken societies. What is necessary, however, is that all stakeholders in the educational system (such as educational institutions, teachers, students, unions, parents, community organizations, private and public sector employers, policymakers, and government agencies) be collaboratively engaged in the process.

Peace education needs to transcend individual person, place, and situation, and seek to incorporate the contextual variables if meaningful change is to become possible. Multidimensional strategies that include all functional aspects of the educational enterprise need to be pursued. Schools need to be run themselves as democratic microsocieties where both students and staff engage in decision making. Beyond the formal curriculum, everything from the pictures on the walls, to the textbooks in use in the classroom, to the admission and testing policies that impact student progress, should speak to the values of equity, mutual respect, and interdependence. Those who educate and lead students should embody the principles of civic responsibility in their ways of knowing and being. Dialogue, inclusion, and participatory methods of pedagogy and problem solving should prevail throughout school communities.

At the end of the day, if peace education is to be a viable vehicle for promoting social cohesion, efforts need to go beyond individually based endeavors and move toward planned, integrated, comprehensive, system-wide action. Only in this way will peace education efforts bear sustainable fruit.

## References

Church, C., Johnson, L.S., and Visser, A. (2004). A path to peace or persistence? The "single-identity" approach to conflict resolution in Northern Ireland. *Conflict Resolution Quarterly*, 21: 273–294.

Constantinou, C. and Papadakis, Y. (2001). The Cypriot state(s) in situ: Cross-ethnic contact and the discourse of recognition. *Global Society*. Retrieved December 22, 2003 from http://www.cyprus-conflict.net/papak-constan.htm.

Douglas, L. (2003). *Citizenship education and human rights education: An overview of recent developments in the UK.* Manchester, England: British Council.

Furlong, J. (1992). Reconstructing professionalism: Ideological struggle in initial teacher training. In M. Arnot and L. Barton (Eds.), *Voicing concerns: Sociological perspectives on contemporary education reforms* (pp. 162–169). Brightwell-Sotwell: Triangle Books.

Gallagher, A., Osborne, R. and Cormack, R. (1993). Community relations, equality and education. In R. Osborne, R. Cormack, and A. Gallagher (Eds.), *After the reforms: Education and policy in northern Ireland* (pp.177–195). Aldershot, England: Avebury.

Gallagher, A. and Smith, A. (2002). Selection, integration and diversity in Northern Ireland. In A.M. Gray, K. Lloyd, P. Devine, G. Robinson, and D. Heenan (Eds.), *Social attitudes in Northern Ireland: The eighth report* (pp. 120–137). London: Pluto.

Hadjipavlou-Trigeorgis, M. (2000). *A partnership between peace education and conflict resolution: The case of Cyprus.* Retrieved December 12, 2003 from http://construct.haifa.ac.il/~cerpe/papers/mariaht.html

Johnson, L.S. (2002). The practice of integrated education in Northern Ireland: The teachers' perspective. Londonderry, NI: UN/INCORE Research monograph series. Retrieved June 29, 2005 from www.incore.ulst.ac.uk/home/publication/occasional/

Johnson, D.W. and Johnson, R.T. (1994). *Leading the cooperative school.* Edina, MN: Interaction Book Company.

Leichty, J. and Clegg, C. (2001) *Moving beyond sectarianism: Religion, conflict and reconciliation in Northern Ireland.* Blackrock CO, Dublin: The Columba Press.

McClenahan, C., Cairns, E., Dunn, S., and Morgan, V. (1996). Intergroup friendships: Integrated and desegregated schools in Northern Ireland. *The Journal of Social Psychology,* 136: 549–558.

McDonald, H. (2002). Catholic stance on integrated schools attacked. *The Observer.* Retrieved January 6, 2002 from www.observer.co.uk/ nireland/story/0,11008,628455,00.html

Montgomery, A., Fraser, G., McGlynn, C., Smith, A., and Gallagher, T. (2003). *Integrated education in Northern Ireland: Integration in practice.* Coleraine: UNESCO Centre, University of Ulster.

Northern Ireland Council for Integrated Education (2004). *NICIE Annual Report March 2004.* Belfast, NI: Northern Ireland Council for Integrated Education.

O'Connor, F. (2002). *A shared childhood: The story of the integrated schools in Northern Ireland.* Belfast, NI: Blackstaff Press.

Phtiaka, H. (2002). Teacher education for a new world. *International Studies in Sociology of Education,* 12: 353–374.

Pingle, F. (1999). *UNESCO guidebook on textbook research and textbook revision.* Paris: Georg Eckert Institute for International Textbook Research.

Robertson, S. (2000) *A class act—Changing teachers' work, the state and globalisation.* London: RoutledgeFalmer.

Smith, A. (2003). Citizenship education in Northern Ireland: Beyond national identity? *Cambridge Journal of Education,* 33: 15–32.

Smith, A. and Robinson, N. (1996). *Education for mutual understanding: The initial statutory years.* Coleraine, NI: University of Ulster Centre for the Study of Conflict.

Tomlinson, H. and Pashiardis, P. (2002). Reform through participation. *Education Journal,* 63, 18.

Trimikliniotis, N. (2003). Mapping discriminatory landscapes in Cyprus: Discrimination and ethnocentrism in education. *The Cyprus Review,* 15: 1–21.

Yusuf, B. (2003, January 27). Teachers take to the barricades. *Newstatesman* (p. 15).

Volkan, V. (1998). The tree model: Psychopolitical dialogues and the promotion of coexistence. In E. Weiner (Ed.), *The handbook of interethnic conflict* (pp. 342–358). New York: Continuum.

Zuzovski, R. (1997). Living together: The impact of the Intifada and the peace negotiations on attitudes toward coexistence of Arab and Jewish pupils in ethnically segregated and mixed schools in Jaffa. *Mediterranean Journal of Educational Studies,* 2: 37–54.

## Chapter Three
## Multiple Realities and the Role of Peace Education in Deep-Rooted Conflicts: The Case of Cyprus

*Maria Hadjipavlou*

### Introduction

> I am very curious to know
> Who was this Greek-Cypriot reading this book?
> He stopped on page 48.
> Perhaps he was called up at that moment?
> And what is more, the title is:
> Man is not born a soldier.
>
> We could have shared memories,
> Eaten ice-cream together,
> I might have dressed the wound on your hand,
> Been able to wear your coat on a wet day.
> I would have liked you to know of my surprise at myself
> —how I can continue with your unfinished book,
> here, like that! . . .
>
> (Yashin, 2000)

After the 1974 events, the coup d'etat and the Turkish invasion of Cyprus we witnessed the massive displacement of people from both communities. Turkish Cypriots were given Greek Cypriot homes and properties to settle in and many Greek Cypriot refugees who fled from the north to the south of the island either stayed initially in refugee camps or were given Turkish Cypriot homes and properties. For many this act of forced political exchange meant "invading" the private space, memories, and family history of the first real owner. Much literature—poetry, prose and drama—was produced on both sides of the divide about this experience and its meaning. Some of it empathetic to the absent other who, through the material objects left behind, gradually becomes present in the new life of the new owner as narrated in the epigraph written by a Turkish Cypriot whose family was given a Greek

Cypriot house. Some other literature selectively excluded the other and mourned the loss of its own side and kept the longing alive with the "Do Not Forget" campaign. This campaign was launched on both sides of the divide. Respective governing elites called upon their own side to "not forget" or rather to remember selectively certain "chosen traumas and chosen glories" (Volkan, 1978). This official appeal that gives rise to different and competing narratives and by extension to different realities contributed in part to the construction of the conflict culture in the past 30 years mainly based on the past and the understanding of the problem as defined by each side separately. Education, being the state institution, that is used to reproduce the official narrative and interpretation of the conflict can also play a significant role in socializing the new generation in peace education and conflict resolution.

Thirty years later on April 23, 2003, the partial opening of the *Green Line* allowed citizens to cross to and from the other side. Thousands of Greek Cypriots, Armenian, Maronities, and Latin Cypriots visited their homes and properties and so did the Turkish Cypriots. Statistics shows us that the first six months (April–October 2003) about one and a half million Greek Cypriots and 700,000 Turkish Cypriots crossed. The euphoria at the grass-roots level was a promising dynamic which should have been translated into political initiatives at the policy level. Many who saw their homes and properties returned with new reconstructed memories, experiences, stories, and messages. The past was recalled and relived by those who owned the houses and properties and those who listened to the memories of the past owners in their present experience. Both shared how they were born in the same house, in the same room but at different times, so each rightly claimed as his/her birthplace. Irrespective of the side to which one belonged, both were afraid of the future. Both felt embodied in a past and both desired a future different from the present that sits on the past. But neither dared to clearly articulate as yet about this future and how it is imagined. Some felt, after a few trips, that it was time to "say goodbye" to what used to be theirs but felt it was no longer so. Some analysts, international journalists, and academics spoke about a new history been made those days. That dynamic connected people who were former "enemies" or invisible, nameless "others." The young generation that had been simply taught through mediated information about the "other side, the fearful other", has now an opportunity to experience it and wonder what would be next.

The Armenian Cypriot poet, Nora Nadjarian (2003), in one of her poems entitled "Don't Forget" gives vividly the intermix of different "pasts" occupying the sleeping and waking life of many Cypriots putting the victim and the "perpetrator" in interchangeable roles as well as transcending these roles to meet in the space of shared pain and wondering who is the "guest" and who the "hostess" in the new present.

> The past came to visit again last night
> Wrapped her arms round my neck
> And whispered: It's me. Don't forget.
> I knocked at a door which a woman opened.
> She said in Turkish: Come in. Welcome.
> Hoggeldiniz Hoggeldiniz.
>
> She handed an album of photos at me,
> My husband, our children this house,
> Pre-1974. The blue album. My living room.

> I kept these for you she said
> I thanked her in Greek. Efcharisto poli.
> A tiny space the size of a pinhead
> Between each word stung the air, the moment.
> The dream. She offered coffee and sweets.
> One of us was guest, the other hostess-but which one?
>
> All those thirty year old tears, finally, belatedly
> two sisters, who were mothers, wives, daughters,
> so long ago. Then the past came and sat between us
> and woke me with a whisper:
> It's me. Don't forget.
>
> (Nadjarian, 2003)

In this chapter I first give a brief history of the Cyprus conflict and second analyze the concept of peace education in protracted conflicts like the Cyprus one. Third, I address the following questions: How can education address the deep divisions in Cyprus and what kind of shared curriculum can be introduced in an effort to move toward reconciliation and shared values? The training of teachers should also be changed in view of the fact that Cyprus, a member of the European Union, aspires toward reunification and the establishment of integrative institutions and shared values. I conclude with some general remarks on the proposed paradigm.

## Cyprus: History of the Conflict

The geostrategic location of Cyprus in the easternmost part of the Mediterranean has made it vulnerable to outside conquests and interference. Whichever power dominated the region controlled Cyprus as well. The Achean Greeks settled on the island in the second millennium BC, formed city-kingdoms on the Minoan model, and introduced the Greek language and culture. To this day the Greek Cypriots refer to this period to stress the Hellenic heritage and its continuity to the present. The Turkish Cypriots stress the three centuries of the Ottoman presence (1571–1878) that determined the interethnic character of the island. After the Ottoman rule, the Moslems who stayed on the island formed what later became the Turkish Cypriot community. By 1960, the Turkish Cypriots comprised 18 percent of the population and Greeks 80 percent, with 2 percent Armenians, Maronites, and Latins.

In 1878, the British took control of the island and Cyprus became a British colony in 1925. During the anticolonial struggle in 1955–59 the Greeks of Cyprus fought the British for "enosis" (union) with "motherland" Greece whereas the Turks of Cyprus demanded "taxim," partition, that is union of part of the island with "motherland" Turkey. The 1950s was a period of intense interethnic mistrust and fears. For instance, according to Turkish Cypriot writers (Nedjatigil, 1997; Salih, 1968), the Turkish Cypriot leadership expected that sooner or later the Greek fighters would terrorize the Turkish Cypriot community, and so by 1957 the Turkish Resistance Organization (TMT) was formed in an effort to counteract the National Organization of Cypriot Fighters (EOKA).

The British politicized communal differences to serve their colonial interests in the Middle East (Pollis, 1998) and reinforced the rise of the two antagonistic nationalisms and competing visions for Cyprus based on each group's "primordial attachments" to their respective motherlands. A compromise settlement was worked out by outside stakeholders—Greece, Turkey, and Britain—that led to the creation in 1960 of the Cyprus Republic (Kyriakides, 1968; Xydis, 1973).

Cyprus is a case of an imposed settlement that ignored local realities (culture of intolerance) and micro-level concerns (Kitromilides, 1977). The imposed accommodation and constitutional arrangements remained fragile and interethnic violence broke out in December 1963 and later in 1967 (Kyriakides, 1968). This resulted in the creation of a "Green line," that is, a dividing line in the capital Nicosia to keep the two warring factions apart. This line was drawn by a British commander and was later patrolled by the United Nations Peace Keeping Force. Turkish Cypriot enclaves were set up in the major cities of the island where Turkish Cypriots moved for security reasons. The Turkish Cypriots having withdrawn from the government the Cyprus Republic was run exclusively by the Greek Cypriots and the Turkish Cypriot vice president declared the constitution "dead."

The period between 1963 and 1974 was a time of unequal social and economic development, a factor that drew the two communities further apart. Greek Cypriots experienced economic prosperity and modernization, whereas Turkish Cypriots entered a period of economic and cultural dependency on Turkey, which they regarded as their "protector" from Greek Cypriot domination. In the early 1970s, extremist groups in both communities with the help of outside parties—the Greek junta—launched a coup d'etat in July 1974, followed by the Turkish invasions on July 20 and August 14, respectively. This led to de facto partition of the island followed by massive displacement of both communities, hundreds were killed, thousands became missing persons, and an economic catastrophe ensued. In 1983, the Turkish Cypriot leadership declared the Turkish Republic of Northern Cyprus (TRNC) but failed to gain recognition by any other country but Turkey, which meant international isolation of the Turkish Cypriot community and increased dependency on Turkey. Over the years the demographic composition in the north has changed with the influx of people from poor Turkish regions. This created what became known as "the settlers' problem."

A series of intercommunal high-level negotiations have been conducted on and off since 1975 under the United Nations auspices, but to this day no agreement has been reached. The latest initiative was the UN "Annan Plan" for a comprehensive settlement but was defeated in referenda whereby the Greek Cypriots voted by 76 percent "No" to the Plan and 67 percent of the Turkish Cypriots voted "Yes." This marked a new turning point in the recent history of the conflict (Bryant, 2004).

As mentioned earlier the partial opening of the Green Line and the crossings of people without any violent incident created a euphoria and a public desire for a solution. The 2003 elections in the north gave a majority to the pro-solution forces headed by a left wing "prime minister and later president." In Turkey the government of Mr. T. Erdogan supported the Annan Plan and advocated settling a date for European Union accession negotiation. On May 1, 2004, the Cyprus Republic (the south of the island) became a member of the European Union. The atmosphere that followed further polarized the feelings on each side of the divide, causing the euphoria and hopes in the Turkish Cypriot community to die off and suspicions and a renewed sense of betrayal to ensue.

In the meantime, new generations have grown up in each community receiving mostly distorted information about the other. Feelings of mistrust, stereotyping, and psychological distancing still abound. Selective histories and memories are used as "text" to dehumanize the other and thus justify the division (Bryant, 1998; Hadjipavlou, 2003). It is within this context that pro-solution groups and individuals from both sides have been challenging the conflict culture. They articulate alternative discourses to the official adversarial ones and aim at building a peace culture based on mutual understanding, respect, and inclusion. They stress *citizenship* as a shared value and a unifying point of reference rather than the traditional ethnic identities historically exploited by the official discourses on both sides, turning them into points of contest and separation. The dualisms, bipolarity, and perceived homogeneity in each community are challenged as oversimplifications of a much more complex social and political landscape. In the case of Cyprus the top-down agreement of 1960 did not work for long due to lack of loyalty and commitment by the people. The failure of the Annan Plan points to the lack of a *culture for* a *solution* as well as lack of linkages between macrolevel and microlevel activities so as to promote the necessary solution mentality (Hadjipavlou-Trigeorgis, 1998). I refer more specifically to this concept later on in the chapter. Let us now examine the concept of peace education.

## The Concept of Peace Education

In protracted conflicts, peace in its broader sense means, according to Kelman (1987, 1990) more than military disengagement and a nonbelligerent agreement. It means a *resolution* of the conflict, an outcome that meets the basic human needs of the parties and is responsive to their basic fears and concerns. Peace from a conflict resolution perspective means an environment conducive to coexistence, cooperative relationships, and reconciliation. This means doing parallel work at both the official and unofficial levels with linkages between them, all aiming in transforming the conflict habituated system (Burton, 1990; Diamond and McDonald, 1991; Hadjipavlou, 1998, 2004; Kelman, 1990; Lederach, 1995).

Peace education according to some peace research scholars looks at the different forms of violence both international and domestic whereby violence in its broader sense includes physical, psychological, and structural that can be caused by thoughts, words, and deeds; in other words any dehumanizing behavior that intentionally hurts another (Harris, 2002). According to Harris, peace educators adapt their approaches to address these different forms of violence within specific, social contexts. In Cyprus, for instance, peace education refers mostly to the efforts undertaken at the unofficial citizens' level comprising educators and students who promote the ideology of rapprochement, and mutual acknowledgment of joint responsibility for the division and pain on the island as well as engaging in joint activities and projects, such as a call to rewrite history textbooks so as to address crude omissions and negative ethnic stereotypes (Hadjipavlou 2003; Morgan 2004). In Ireland peace education refers to their program on "education for mutual understanding," cross-community contacts and the establishment of integrated schools (Cairns, 1987; Duffy, 2000).

Peace education, in addition to dealing with curriculum materials is also a tool, a skill as much as a mental attitude, a cultural construct. In the United States of

America and Europe during the 1980s much of peace education focused on the dangers of the "nuclear holocaust" and most recently in the 1990s peace education became connected to conflict resolution and how to manage conflicts in a nonviolent way both in the classroom and outside.

In the United States of America, Europe, and elsewhere "there are currently more than 300 colleges and universities offering peace studies courses, and with more than 150 having programs" (Harris, 2002, p. 24). The students and educators involved in peace studies recognize that both the material and spiritual survival of the human race demand education for peace. Many studies show the extent to which people have been educated to expect war and not peace and this needs to change. (Forsey, 1989). Many researchers and intellectuals have proposed ways to "learn how to change our thinking through peace studies" (Boulding, 1989; Diamond, 2001; Fisher, 1994; Forsey, 1989; Kant, 1975; Montessori, 1974; Salomon, 2002; Wallersteen and Axell, 1994).

In recent years we note an increased academic interest and research in peace education due to the fact that whole societies have been experiencing violence, armed strife, division, economic hardships, insecurity, and fear of eliminations and flagrant violations of human rights. Such places are Palestine and Israel, Rwanda, Northern Ireland, Cyprus, Bosnia, Kosovo, Afghanistan, Iraq, and so on. Many of us expected that after the end of cold war the world would be safer and more global cooperation and the realization of interdependence would prevail. Some of us had hoped that much more funding would go to issues of education, health, and environmental security. Instead, toward the end of the twentieth century we have more than 32 ongoing armed conflicts, many more refugees, dead, displaced persons and an increase in terrorist activities (Wallersteen and Axell, 1994). We still live in a militarized, patriarchal and gendered world where the military expenditure per soldier is 25 times higher than the money spent on education for each child in the world (Sivard, 1993).

Like violence, peace has different manifestations and various scholars have given us different typologies or categorizations. For instance, some speak about inner and outer peace, negative and positive peace, structural peace, ecofeminist peace, intercultural peace and so on. Peace education according to Ian Harris (2002) attempts to draw out of people their natural inclination to live in peace: "Peace researchers identify processes that promote peace, whereas peace educators, educating people about processes, use teaching skills to build a peace culture" (p. 18). Moreover, Salomon (2002) informs us that peace education in regions of intractable conflicts often includes elements of antiracism, conflict resolution, multiculturalism, cross-cultural training, and the cultivation of a generally peaceful outlook. One of the most difficult obstacles that peace educators face refers to collective animosities, shared painful memories, and commonly held view of self and other, issues that should also be of concern to the macrolevel as well:

> Peace education in regions of intractable conflicts uniquely confronts what Azar has described as "ethnic [racial or religious] hostilities crossed with developmental inequalities that have a long history and bleak future." It follows from this conception that this class of peace education faces three important challenges: (1) it faces a conflict that is between collectives, not between the individuals, (2) it faces a conflict that is deeply rooted in collective narratives that entail a long and painful shared memory of the past and (3) it faces a conflict that entails grave inequalities. (Salomon 2002, p. 7)

Salomon further outlines a conception of Peace Education in the context of intractable and protracted conflicts (Azar, 1985; Bar-Tal, 2000). He proposes *four* kinds of interrelated dispositional outcomes. It is assumed that the collective historical memories and narratives each conflicting side holds about the conflict affects the views and attitudes of individuals on the other side; how they relate to the other as well as how they each interpret the actions of the others. This mental frame leads to the dichotomy of "us versus them" of mutual victimhood and feeds misperceptions and the perpetuation of the conflict. One of the interventions peace education can make is at the level of changing perceptions about the self and other and the others' collective narrative.

The first "dispositional outcome" refers to the *legitimation of the Other's Collective Narrative* that means focusing on the conflict as defined between collectives and not between individuals, for instance, the enemy is neither Ahmet nor Eleni but governments and institutions. When we accept the other's narrative it means that events, past and present, can be seen from both lenses. Moreover, by legitimating the other's narrative as much as one's own, a space is opened for dialogue and trust building, necessary ingredients for the second disposition that refers to a *critical examination of our contribution to the conflict*. This means that the suffering, which each side has inflicted on the other, is acknowledged together with the guilt and denial that usually accompany such avoidance. In the conflict resolution workshops among groups of Greek and Turkish Cypriots we promote this disposition and by the end we note a shared narrative based on a new reality emerging liberating both sides from their "egoism of victimhood" as well as from fear of how they will be perceived when they reveal their own "badness."

The third disposition *refers to empathy for their suffering*. In conflict situation each side gets wrapped up in its own suffering, pain, loss, and trauma. Developing the skill of empathy means the ability to appreciate and feel with the other's pain, agony, and loss and thus engage in a process of mutual humanization. Many activities and conflict resolution workshops in Cyprus have focused on developing this disposition because decades of separation reinforced each side's mistrust and anger.

Finally, the fourth disposition or goal of peace education is *the engagement in nonviolent activities*. This means developing a consciousness that violence is not the way to solve social or international conflicts and instead use the tools of dialogue, negotiations, confidence building through contacts, and so on. This level is more skill-oriented.

### An Integrative Pluralistic Paradigm: Underlying Principles

Let us now consider the question as to how new generations can be educated in a way that can both *respect* each other's cultural identity and *prepare* them for citizenship in a future democratic, reunited, federal, multicultural Cyprus. What are the principles that would help us build new *dispositions* as outlined toward mutual acceptance and a culture of peace education in view of all the competing histories and realities that still prevail in the minds and experiences of Cypriots ? What I present is partially based on the views and recommendations expressed by hundreds of Cypriots in bicommunal dialogues, structured meetings, and conflict resolution training workshops over a number of years in which I was a participant, a participant observer, or a facilitator. I believe there exists by now a number of core support groups of citizens, educators, students, and other professionals who are ready to commit to a new way of thinking in order to socialize the future Cypriot citizen in the values of democracy, pluralism, nonviolence, and multiculturalism (see www. hisdialresearch.org).

The proposed paradigm is characterized by (1) *inclusion* (a big shift from the "either . . . or" mode that currently prevails) this also means developing a disposition whereby the collective narrative of the other is legitimated; (2) *flexibility* that is enlarging the curriculum to give recognition to the part that each side has been excluding or silencing, something that will lead to both self and other understanding as well as constructing a shared narrative for coexistence; (3) *openness* to dialogue, which means diversity of interpretations and worldviews, recognition of multiple realities and constant critical assessment of one's social, educational, and political system so as to keep both the system alive and democracy dynamic; (4) *mutual humanization*, which means acknowledging mutual hurts and being sensitive to each side's historical grievances, traumas, and values as this relates to the Salomon model of acknowledging one's own contribuition to the conflict and promoting the ability for empathy of the other's suffering and pain. My research (2003) has shown that each community considers its own side and the errors committed by its leadership to have contributed to the creation of the conflict and having inflicted pain on the other (87.1 percent of Greek Cypriots and 65 percent of Turkish Cypriots believe so). This is a hopeful shift that can be utilized to introduce a complex view of self and other and the undertaking of responsibility; (e) *mutual awareness* that is cooperation recognition of interdependence. In other words, the challenge is how to educate the Cypriot citizen of tomorrow in participatory democracy and in a multicommunal Cyprus where differences would be appreciated. What are the prerequisites that would help create a culture of peace?

It would help both communities to keep separate the ethnic/cultural identities from their political identities that means shifting political loyalties from their referent nations (Greece and Turkey) to the future unified, federal, and democratic Cypriot state so as to be able to share interethnic interests in an open society. My research (2003) on the issue of collective identity has shown that high percentages define themselves as "Cypriot" first in a scale of various options (Hadjipavlou, 2003). This is indeed a hopeful shift in the sense that people now relate more to the land and geograpy of Cyprus than to "motherlands." Historically, politicization of cultural/ethnic loyalties have been a threat to civic identity for a common citizenship. Thus social criticism and serious reflection on social values relevant to the needs of a conflict-ridden society are fundamental. It is important to lift the pressure from the future Cypriot citizens to have to choose between different parts of their identity, European, Cypriot, Greek, Turkish, Armenian, Maronite or Latin, or any others. The new citizenship should be built on relatedness and in conversation with rather than competition and domination of one group over the other. Identities are not fixed givens but in a state of constant transformation.

Ultimately, what is involved in multicultural and peace education in Cyprus is much the same as that that is involved in the development of a democratic public whereby the process of understanding one's own culture is as complex as is the process of understanding the other's culture. This means to view one's own activity and behavior as a cultural product and to avoid defining it as modeling cultural norms. Thus, a major task for peace education in a democratic, multicommunal society is coming to terms with "otherness" (Hadjipavlou, 2002). To this end the use of a shared curriculum in divided societies would help.

## Shared Curriculum

The introduction of a common *Civics* curriculum, for example, will help foster the Cypriots' understanding of citizenship responsibilities and rights and would promote the principles of pluralism and the need to be taught something about the other fellow-citizens' culture and creative expression. At present students learn more about Greece and Turkey than about Cyprus. This will need to change within the European Union context. Students would be able to develop a knowledge and understanding of the similarities and differences between the cultural traditions and historical experiences that influence people in their country. Greek Cypriot students would also learn about the Turkish Cypriot culture (language, religion, customs) and traditions, the Armenian, Latin, and Maronite ways of life, without fearing that this threatens their uniqueness as Greeks and Cypriots. The same applies to all other ethnic groups. It would also be educationally enriching to produce television and radio programs about the different cultures and their symbiosis on the island. A multicommunal journal where creative minds from all Cypriot communities can publish their different experiences emanating from the same homeland. Using technology new series of public education can begin introducing the societies to the complexity of experiences and practices. Gradually, a new language and positive images will be created ensuring that different identities can coexist and that the Cypriot citizenship can act as a unifying force.

Another common curriculum course should aim to expose all the Cypriot students to the *system* and *ideals* of *democracy*, its different kinds, its own tensions and strain as well as raising consciousness against any form of discrimination, promoting support for human rights for all. After all, democracy is a way of life and cannot be the exclusive possession of any ethnic group, or social class, or any subgroup of a body of citizens. I believe this is crucially important for Cyprus because we often witness the resignation of people from politics and civic involvement, a sign that participatory democracy becomes the exclusivity of the few elites.

Living in a militaristic, patriarchal, and hierarchical society as recent research on women in all Cypriot communities has shown (Hadjipavlou, 2004) it is appropriate to introduce early on *Gender and Peace Studies and Human Relations* courses so as to sensitize students to alternative worldviews and the consequences of the social construction of gender. The impact of the conflict and separation on men and women, young and old, in urban and rural regions, as well as studying the causes of conflict need to be compulsory subjects so as to build an antimilitaristic, peaceful, gender equality, and humane culture.

Related to this is the teaching and practice in *conflict resolution* skills, problem-solving, and *communication* skills. Some of my male university students often tell me that in both the secondary schooling and in the army they were exposed to a hostile disposition toward the Turks and even believed in killing them:

> It was my duty to kill a Turk because they (Turks) unjustly occupied our land and deprived us of our homes and properties. And I grew up believing in this. When the Green Line opened and Turkish Cypriots started coming to our side I was scared that they would harm us and dirty our side too. Now that I see this did not happen I feel differently. But for many years I was made to believe that I had to stay away from them or fight against them.

Such misperceptions and negative stereotypes need to be addressed with more research on each side. In a bicommunal research study on ethnic stereotypes we noted that the Greek Cypriots attributed all the positive characteristics to themselves and the negative ones to the Turkish Cypriots whereas the Turkish Cypriots held a more complex view of the Greek Cypriots, both positive and negative (Hadjipavlou, 2003; Spyrou, 2002).

In another study (2004) on "women in all Cypriot communities" the smaller communities, Maronites, Armenians, and Latins spoke about facing cultural extinction and discrimination. They mainly explained it on the grounds of exclusive education as it is framed and promoted by the dominant community, that is, the Greek Cypriot ministry of education. They complained that there is nothing in the text books-history literature or geography about their community, their origin on the island, their culture or needs, and their special contribution to the overall Cypriot culture and economy. This has to change especially in view of the European standards and principles of minotity rights (Hadjipavlou, 2004).

## Teacher Training

The above ideas and goals of peace education can only materialize by enlarging the teachers' education curriculum and training and building linkages with parents associations too—the family is still a strong institution in Cyprus. Teachers should be exposed to new resource programs for skill training—communication, mediation, negotiations. Students should undertake joint projects on social studies, gender studies, and local history issues and thus build partnerships across ethnic and gender lines. Already this started happening at an individual and voluntary level through the guidance of peace builders who are educators themselves. Some of them formed nongovernmental organizations for dialogue on the multiple (his)stories and are researching how official history is taught in both sides.

In the Turkish Cypriot community an initiative resulted recently in new history textbooks. This should have been a joint, collaborative effort between educators from all Cypriot communities. In the Greek Cypriot side where the problem of prejudice and racism among the youth is widespread there are no official initiatives to address these issues in a systematic way. At the unofficial level there are so far close to 1200 youth, boys and girls who have gone through conflict resolution workshops and activities of the Youth Encounters for Peace (YEP) project. This is a voluntary interethnic project that started in 1997 led by two inspiring teachers across the divide (see: www.humanrights-edu-cy.org) and aims at building mutual understanding and mutual acceptance through contacts.

The institution of education can become a support system for conflict transformation, peace education, and future coexistence. Educators and students in the conflict resolution encounters have expressed the need for increased and open communication among the new generation; they want to build trust and get rid of the mutual fears; emphasize commonalities and past shared experiences; institute a process to heal old wounds and get rid of the "enemy image"; build strong, democratic civil and political institutions and bicommunal organizations; be open to new ideas and assume citizen responsibility to do something for peace. The youth's desire for peace can impact change at the macro level. These dispositions are in concert with the Salomon model of peace education as outlined earlier.

With regard to undertaking activities and joint projects, many educators and students in their cross-ethnic encounters have proposed the organization of peace camps for youth from both communities; establish a Cypriot publishing house to publish new joint materials and books (e.g., Cypriot fairy tales, oral histories, local practices, and shared customs) and disseminate this new resource in all schools so as to complement prescribed curriculum. The establishment of a multicommunal research center where interested scholars from all Cypriot communities can engage in joint research projects (hopefully funded by EU) would develop new knowledges. Learning each other's language and literatures would enrich appreciation and empathy building.

Since 2003 journalists and others have contributed articles in each other's newspapers supporting the ideology of rapprochement. These articles can constitute complementary reading in schools for it allows a different view of the current realities to be discussed from both perspectives. Of course, the establishment of a Peace Radio or TV Stations to disseminate the voice and language for coexistence and cooperation would enrich further the principle of multiculturalism and reconciliation at multiple levels. In addition, this activity would contribute toward the creation of *a culture for a solution.*

What does a "culture for a solution" mean in a divided country? Briefly it means a change in the conflict mentality, a resocialization processes that encompasses all four dispositional steps as given by Salomon (2002). In addition we need a well-designed policy adopted by both state institutions and nongovernmental organizations and influential individuals so as to initiate a multiple point of entry into the conflict habituated system (Diamond and MacDonald 1991). All the above propositions entail changes in the adversarial mindset and in public discourses that emanate from the officials, the community leaders, the mass media, and some nongovernmental organizations. We thus need a multitrack intervention whereby the education system, its philosophy and content to include the other and multiple perspectives of the historical events would be a priority. The role of the church in the Greek Cypriot community in influencing people should change to promote understanding, reconciliation, inclusion, empathy, and coexistence.

New images in the form of monuments and statues, parks celebrating peace, and joint struggles for freedom should be constructed replacing the war monuments reifying nationalism and exclusion. One of the findings in a recent research contacted in all the communities concerning the creation and perpetuation of the Cyprus conflict high percentages of Greek and Turkish Cypriots (69 percent and 66.5 percent respectively) mentioned the different values and beliefs cultivated by the separate educational systems. Even higher percentages (74.9 percent and 87.7 percent) believed that the lack of trust and communication contributed to the perpetuation of the conflict as much as local nationalisms (see Hadjipavlou, 2003, 2004B). Thus the introduction of sustained peace education is long overdue in Cyprus.

## Concluding Remarks

For historical and political reasons the prevalent education paradigm in the two communities has been based on the "us versus them" dichotomy and students are educated and socialized in the conflict. The psychological and geographic separation of teachers and students has been challenged in the 1990s through the systematic use of conflict resolution training workshops and interethnic encounters and seminars that expose citizens to alternative ways of viewing each other, and provide opportunities to address underlying fears, concerns and basic needs of each community.

The proposed *integrative, pluralistic* educational paradigm calls for abandoning the competitive and antagonistic approach to addressing differences and conflict experiences using a new joint problem solving–oriented paradigm that promotes peace education based on the principles of democracy, pluralism, and multiculturalism. Peace education can thus function as a healing platform for past mutual grievances and develop joint responsibility toward a shared future. A new inclusive and more complex narrative should thus be born out of this educational peace vision in a new Cyprus where education is linked very appropriately to the bigger reconciliation efforts on the island whereby multiple realities constantly unfold.

The Cypriot intellectuals who can imagine a different Cyprus, together with NGOs can exert pressure on the leadership and ministers of education in both Cypriot communities and other interested third parties to listen to the encouraging messages that peace builders and youth from all Cypriot communities have been sending. Their message is one of "a desire for a normal life." The role of intellectuals is dual: to promote further research in all communities and produce data on the current realities and use scientific analysis for informing policymakers; and to engage in an alternative social discourse where self-criticism and the capacity for empathy and undertaking of own responsibility become tools in enlarging the mental and emotive maps of all citizens. Hopefully this effort will lead to a state of shared joys of coexistence and creative tensions instead of fear of violence that will deepen separation and alienation.

## Note

I would like to thank Zvi Bekerman and Claire McGlynn as well as three reviewers for their constructive comments and suggestions in improving this text.

## References

Azar, E.E. (1985). Protracted social conflict: Ten propositions. *International Interactions*, 12: 59–70.
Bar-Tal, D. (2000). From intractable conflict through conflict resolution to reconciliation: Psychological analysis. *Political Psychology*, 21 (2): 351–365.
Boulding, E. (1989). Can peace be imagined? In L.R. Forcey (Ed.), *Peace, meanings, politics, strategies* (pp. 73–86). New York: Praeger

Bryant, R. (1998). An education in honour: Patriotism and the schools of Cyprus. In V. Calotychos (Ed.), *Cyprus and its people: Nation, identity, and experience in an unimaginable community 1955–1997* (pp. 53–68). Boulder, CO: Westview Press.

———. (2004). An ironic result in Cyprus. Middle East Report online. Retrieved from www.merip.org. May, 2004

Burton, W.J. (1990). *Conflict: Human needs theory.* New York: St. Martin's Press.

Cairns, E. (1987). *Caught in crossfire: Children and the Northern Ireland conflict.* Belfast and Syracuse, NY: Appletree Press and Syracuse University Press.

Diamond, L. (2001). *The peace book.* Berkley, CA: Conari Press.

Diamond, L. and McDonald, J. (1991). Multi-track diplomacy: A system's guide and analysis. Iowa Peace Institute Occasional Paper 3. Grinnell, Iowa: Iowa Peace Institute.

Duffy, T. (2000, March). Peace education in divided society: Creating a culture of peace in Northern Ireland. *Prospects: Quarterly Review of Comparative Education,* 30 (1): 15–20.

———. (1994). *Interactive conflict resolution.* Syracuse, NY: Syracuse University Press.

Forsey, L.R. (1989). Introduction to Peace Studies. In R.L. Forcey (Ed.), *Peace, meanings, politics, strategies* (pp. 73–86). New York: Praeger.

Hadjipavlou, M. (2002). Cyprus: A partnership between conflict resolution and peace education. In G. Salomon and B. Nevo (Eds.), *Peace education, the concept, principles and practice around the world* (pp. 193–209). New Jersey and London: Lawrence Erlbaum Associates Publishers.

———. (2003). Inter-ethnic stereotypes, neighborliness, and separation: Paradoxes and challenges in Cyprus. *Journal of Mediterranean Studies,* 13 (2): 340–360.

———. (2004a). The contribution of bicommunal contacts in building a civil society in Cyprus. In A.H. Eagly, R.M. Baron, and V.L. Hamilton (Eds.), *The social psychology of group identity and social conflict: Theory, application and practice* (pp. 193–213). Washington, DC: APA

———. (2004b). *Women in the Cypriot communities.* Nicosia, Cyprus: PC Press

Hadjipavlou-Trigeorgis, M. (1998). Different relationships to the land: Personal narratives, political implications and future possibilities. In V. Calotychos (Ed.), *Cyprus and its people, nation, identity, and experience in an unimaginable community 1955–1997* (pp. 251–277). Boulder: Westview Press.

Harris, I. (2002). Conceptual underpinnings of peace education. In G. Salomon and B. Nevo (Eds.), *Peace education, the concept, principles and practice around the world* (pp. 15–26). New Jersey and London: Lawrence Erlbaum Publishers.

Kant, J. (1975). Perpetual peace: A philosophical sketch. In H. Reiss (Ed.), *Kant's political writings* (pp. 53–71). Cambridge: Cambridge University Press.

Kelman, H.C. (1987). The political psychology of the Israeli-Palestinian conflict: How to overcome the barriers to a negotiated solution. *Political Psychology,* 8 (3): 347–363.

———. (1990). Applying a human needs perspective to the practice of conflict resolution: The Israeli-Palestinian case. In J. Burton (Ed.), *Conflict: Human needs theory* (pp. 123–145). New York: St. Martin's Press.

Kitromilides, P. (1977). From coexistence to confrontation: The dynamics of ethnic conflict in Cyprus. In M. Attalides (Ed.), *Cyprus Reviewed* (pp. 143–186). Nicosia: The Jus Cypri Association.

Kyriakides, S. (1968). *Cyprus: Constitutionalism and crisis government.* Philadelphia: University of Pennsylvania Press.

Lederach, J.P. (1995). *Preparing for peace, conflict transformation across Cultures.* Syracuse, NY: Syracuse University Press.

Montessori, M. (1974). *Education for a new world.* Thiruvanmiyur, India: Kalashetra.

Morgan, T. (2004, July 19). Revolution in the classroom after decades of hatred. *Financial Times,* p. 23.

Nadjarian, N. (2003). *Cleft in Twain.* Nicosia: Cassoulides and Sons Ltd.

Nedjatigil, Z.M. (1997). *Cyprus: Constitutional proposals and developments.* Lefkoshia: Turkish Federated State Press.

Pollis, A. (1998). The role of foreign powers in structuring ethnicity and ethnic conflict. In V. Calotychos (Ed.), *Cyprus and its people, nation, identity, and experience in an unimaginable community 1955–1997* (pp. 87–105). Boulder, CO: Westview Press.

Salih, H.I. (1968). *Cyprus: Analysis of Cypriot political discord.* New York: Theo Gaus Sons.

Salomon, G. (2002). The nature of peace education: Not all programs are created equal. In G. Salomon and B. Nevo (Eds.), *Peace education, the concept, principles and practice around the world* (pp. 3–13). New Jersey and London: Lawrence Erlbaum Associates Publishers.

Sivard, R.M. (1993). *World military and social expenditure.* Washington, DC: World Priorities Institute.

Spyrou, S. (2002). Images of "the Other": The Turk in Greek Cypriot children's imaginations. *Race, Ethnicity & Education* 5 (3): 255–272.

Volkan, V. (1978). *Cyprus: War and adaptation: A psychoanalytic history of two ethnic groups in Conflict.* Virginia: University Press.

Wallensteen, P. and Axell, K. (1994). Conflict resolution at the end of Cold War, 1983–1993. *Journal of Peace Research* 31: 333–349.

Xydis, S.G. (1973). *Cyprus: Reluctant Republic.* The Hague: Mouton.

Yashin, M. (2000). Don't go back to Kyrenia. Middlesex, London: Middlesex University Press.

# Chapter Four
# Reconciliation and Peace in Education in South Africa: The Constitutional Framework and Practical Manifestation in School Education

*Elmene Bray and Rika Joubert*

### Introduction

South Africa's history has been blemished by racial discrimination and political violence. In the education sphere, a white minority regime forced upon the majority of South Africans (i.e., blacks) an inferior and discriminatory "Bantu" education system that eventually resulted in schools and universities being turned into bloody battlefields.

During the early 1990s leaders across the political spectrum opted for peaceful democratic transformation that culminated in the establishment of a negotiation process for democratic change in South Africa (De Villiers, 1994, p. 1–11).[1] The constitutional negotiations proceeded in a peaceful manner and allowed all the major political parties, interest groups, business and commerce establishments, and individuals to participate, directly or indirectly, in debating a democratic future for South Africa.[2] This process paved way for the drafting and adoption of the 1993 (interim) constitution[3] and its successor, the 1996 constitution (South Africa, 1996(a)). The constitution abolished the previous apartheid system and constituted a sovereign democratic state founded on the fundamental values of human dignity, equality and the advancement of human rights and freedoms, nonracialism and nonsexism, to name but a few.[4]

This chapter explores the constitutional framework and the norms and values it provides for peace and reconciliation in South Africa. The impact of these norms and values in cultivating peace and reconciliation in education are also highlighted. The second part of this chapter highlights the progress made thus far in promoting peace in education, mainly through the fostering of tolerance and respect for difference in a multicultural school environment. Finally, some of the major challenges facing education stakeholders are addressed.

## Constitutional Imperatives for Reconciliation and Peace

The constitution is the supreme law of the Republic. All laws or conduct inconsistent with it are invalid (South Africa, 1996(a): s 2). It means that all law (including education legislation) must be consistent with the constitution and will be tested by the courts for validity. Similarly, conduct (including the one by education officials and educators) must conform to the constitution and will be tested against its provisions for validity. Furthermore, constitutional obligations must be fulfilled, that is, for example, educators must properly fulfill their constitutional obligations toward learners (i.e., respect human rights and promote tolerance in education).[5]

## Constitutional Values and Norms for Achieving Reconciliation, Peace, and Equality

The constitution is a "value-laden" and "value-driven" document that enshrines values that have their origin in the history and experience of the South African people (Botha, 1994, p. 233). They reflect a reaction to authoritarianism and racial exclusivity, and express values and aspirations that would guide us toward a united, democratic nation with a strong value system and human rights culture (Mureinik, 1994, p. 31). These values draw their strength from the formal law-making process, but also from social, moral, and ethical issues that are interwoven with the law: for example, questions relating to dignity and respect for people (*ubuntu*),[6] tolerance and acceptance of diversity, reconciliation, openness and accountability, to name but a few (Rautenbach and Malherbe, 1999, p.11–12). By enshrining these values in the constitution, the people and lawmakers have accepted the paramountcy of these values and are bound by them for the lawful exercise of their powers in terms of the constitution (Botha, 1994, p. 234; Devenish, 1999, pp. 9–17).

Reconciling and reconstructing the South African nation is crucial for peace in the country.[7] Preliminary work in this regard had been done during the constitutional negotiation process that brought a peaceful constitutional solution in South Africa: leaders across the political spectrum undertook to abandon violence and work toward peace and reconciliation (De Villiers, 1994, p. 6).[8] A culture of negotiation was established that enabled and strengthened the development of trust between the negotiators and their alliances and this proved to be invaluable in casting the foundation for reconciliation in a deeply divided society. It also provided the building blocks for the reconstruction of a democratic South African nation, united in its diversity (De Villiers, 1994, p. 1–11). In the spirit of realizing peace and reconciliation, the Truth and Reconciliation Commission (TRC) was later established to, inter alia, assemble a picture of past atrocities, facilitate an amnesty process, make recommendations regarding measures to be taken with regard to victims of human rights violations and propose ways to prevent any future recurrence of human rights violations.[9] Its main objective was thus to promote national unity and bring about reconciliation among the people of South Africa. Despite criticism, the TRC unveiled that a new democratic order could not be achieved without revealing the truth about the deep emotions and indefensible inequities that shrouded the shameful period of apartheid (Rakate, 2004, p. 290–295). Only through this process of finding the truth and obtaining relief from the burden of guilt or anxiety could a commitment to reconciliation and reconstruction be established (*AZAPO v President of the Republic of South Africa*, 1996: paras 2 and 17).

Human rights violations, especially the violation of a person's human dignity (e.g., segregation and humiliation) and a denial of equality (e.g., discrimination on the basis of race and sex) thus constituted some of the worst forms of violations and atrocities during the apartheid years and, consequently, were the rationale behind the political violence and unrest in the country and in educational institutions. Human dignity, the achievement of equality and the advancement of human rights and freedoms, nonracialism and nonsexism are therefore important (and core) constitutional values[10] upon which the new constitutional democracy is built and it is essential that these values and human rights be addressed in the context of peace and reconciliation in education. The values of human dignity, equality, and freedom are also recognized as individual fundamental rights and thus enshrined in the Bill of Rights as individual fundamental rights (De Waal, Currie, and Erasmus, 2001).[11]

The Bill of Rights protects the right of every person to human dignity, equality, and freedom: it thus also guarantees the human dignity, equality, and freedom of every learner in the school environment.[12] Every person has the right to a basic education (South Africa, 1996(a): s 29(1)). This right places a corresponding duty on the state to provide facilities and resources for the provision of such education (South Africa, 1996(a): ss 2, 7, and 237). Children's rights are protected additionally, for example, a "child's best interests" are "of paramount importance in every matter concerning the child"[13]—a principle diligently upheld and enforced by the courts in various court cases since 1994.[14]

The right to equality is a complex cross-cutting right[15] and manifests in many spheres of life, particularly in cases where discrimination based on race, sex, gender, culture, language, or religion occurs.[16] Its inclusion as a constitutional value and an individual human right portrays a deliberate break from the apartheid order that was based on discrimination and the denial of equality.[17] This right must be understood and interpreted contextually, namely, the actual social and economic conditions of groups or individuals have to be considered to determine whether the commitment to equality is being upheld. (Albertyn and Kentridge, 1994, pp.149–150). The right to equality must ultimately ensure equality of outcome.[18] The prohibition on racial discrimination in schools forms part of this right and is reinforced and given effect to in education legislation, including the school's code of conduct.[19]

The constitution provides "checks and balances" to ensure that constitutional values and human rights are enforced. For example, the courts administer justice and are seen as the watchdogs of democracy. They apply the law impartially and without fear, subject only to the constitution (South Africa, 1996(b): s 165). Courts, tribunals, and forums (including disciplinary committees) are instructed to interpret fundamental rights in the light of the values that underlie an open and democratic society based on human dignity, equality, and freedom (South Africa, 1996(a): s 39(1)). Of particular importance is the constitutional court as the highest court in constitutional (including human rights) matters.[20] Provision is also made for state institutions supporting constitutional democracy, such as, the Public Protector, the Human Rights Commission, the Commission for the Promotion and Protection of the Rights of Cultural, Religious and Linguistic Communities, and the Commission of Gender Equality (South Africa, 1996(a): s 181–182). These institutions are not courts but are empowered to act independently, and conduct investigations and submit reports to parliament.[21]

The constitutional values and human rights discussed have permeated the education system and are indispensable for creating peace in education through the cultivation of a human rights culture of tolerance and respect (Ministry of Education, 2001, p. iii–iv). The education manifesto on values, education, and democracy recognizes that values that transcend language and culture are the common currency that makes life meaningful (Ministry of Education, 2001, p. iv). It embraces values such as democracy, social justice and equity, equality, nonracism and nonsexism, human dignity, an open society, accountability, the rule of law, respect and, above all, reconciliation. This document has formed the basis of a training program for educators and has promoted the integration of constitutional values and human rights into the curriculum in all the provinces in South Africa.

## Current Strategies toward Reconciliation and Peace in Education in South Africa

### Towards Peace in Education

Although education for peace is not officially described, there is an increasing awareness of this movement in South Africa (Carl, 1994, p. 11). Peace educators should be clear on what is understood under this concept. Peace is much more than the absence of war. A distinction can be made between negative peace and positive peace. Negative peace implies the ending of direct violence. Positive peace is the eradication of the historical and structural causes for poverty, racism, prejudices, social injustices, and inequality. Education for peace should be a lifelong process developing appropriate knowledge, skills, attitudes, and values to pursue positive peace. The "roots of peace" (Carl, 1994, p. 11) should also be addressed, that is tolerance, cooperation and consensus, a good judicial system, and a democratic government.

Education for peace is a process that tries to solve conflict by the development of certain attitudes, skills, and knowledge. The purpose is to enable children and educators to accept their responsibilities to live in a peaceful, independent, and responsible manner in a diverse community that is often characterized by violence. An essential aspect of peace education is therefore not only the acquisition of knowledge, but of certain important attitudes and values (Ministry of Education, 2001, p. 11).

### Cultivating Respect for Diversity

South Africa consists of communities that manifest different forms of cultural diversity. Diversity in the sense of being different is in itself neither good nor bad although problems arise in relation to the way in which difference is handled. It is, however, normal that a certain degree of tension among competing cultural (read gender, language, age, religious) groups will occur. In fact, at any level of human contact, even within one family of siblings, a certain amount of healthy competition is introduced. There are different environmental demands and, of course, different ethnic, linguistic, gender, and religious characteristics that may cause tension in any relationship.

Education exposes learners to the hard social facts of life where they are taught how to view things in a certain manner and to react to them in a particular way. Dealing with issues of inclusion and exclusion within the framework of the constitution requires understanding and acceptance of differences among South African people who have been forcibly segregated and, consequently, isolated from each other for more than 300 years. True reconciliation not only requires bringing people together at school level but also needs constant compromising and agreements throughout life (Soudien, Carrim, and Sayed, 2004).

The Minister of Education's Tirisano Program

Postapartheid education reconstruction has been a central part of South Africa's overall constitutional and political reconstruction and development. It has been driven by two imperatives: first, the government had to overcome the devastation of apartheid, and provide a system of education that would build democracy, human dignity, equality, and social justice (Asmal, 2001, p. i); second, a system of lifelong learning had to be established to enable South Africans to respond to the economic and social challenges of the twenty-first century. The former minister of education responded with a "Call to Action" using the slogan "Tirisano" that, in the Sotho language, means working together. The Tirisano program focuses on five core programs to transform the education system; these are HIV/AIDS, school effectiveness, literacy, further education and higher education, and organizational effectiveness of the departments of education.

**National Curriculum**

South Africa's democratic government inherited a fragmented and divided education system. Previously, South Africa had 19 different educational departments separated by race, geography, and ideology. In each of these 19 systems learners were prepared in a different way for the positions they were expected to occupy in social, economic, and political life under apartheid. In each, the curriculum played a powerful role in reinforcing inequality.

Curriculum change in postapartheid South Africa started immediately after the 1994 elections. The Revised National Curriculum envisages the infusion of the values of human dignity, social justice, equity, and democracy across the curriculum (Department of Education, 2000). The social sciences learning area, for example, involves the study of relationships between people and between people and the environment at various times and various places. These relationships have social, political, economic, and environmental dimensions. The curriculum aims to develop informed, critical, and responsible citizens who are able to participate constructively in a culturally diverse and changing South African society.

Decentralization of School Governance

In order to lift education out of its divided and discriminatory past, the South African government developed a comprehensive and complex policy framework.

Central to this framework was the idea of devolving power and authority—essentially a decentralization approach—to the stakeholders in education at ground level (Department of Education, 2004, p. 21). Basic to such reform of school governance is the South African Schools Act 84 of 1996 (South Africa, 1996(b)).

The Schools Act provides for the establishment of a single, unified, and uniform system of school education and the establishment of school governing bodies in all public schools. A democratically elected school governing body is composed of parents, educators, and noneducators at the school, and learners in Grade 8 or higher in a secondary school. The phenomenon of self-governing public schools in South Africa is now nine years old and in these nine years much has been achieved. Almost every one of the approximately 27,000 schools has in place a governing body (Department of Education, 2004) of which the main functions include the development of an admission, a language and religious policy for the school, assisting professional staff with learner discipline, and the management of school finances. A school governing performs all its functions in the best interests of the school and strives to develop the school through quality education for all learners at the school. It also administers and controls the school's property and may apply additionally to maintain, for example, the grounds and buildings and provide textbooks. Members of the governing body recommend to the Department of Education the appointment of educators and other staff (South Africa, 1996(b): ss 6–7, 20–21).

### Challenges for Schools, Educators, and Learners in South Africa

The legacy of racial domination has posed not only many challenges but also real and immediate problems that have to be addressed to secure safe school environments and a peaceful education process. Dealing with diversity in the classroom, solving conflict caused by cultural differences, cultivating respect for diversity and differences,[22] the accommodation of various religions (i.e., freedom of religion) within multifaith schools, the balancing of equality of treatment with freedom to be different or to be oneself, and the choice of own schools are some of the practical problems faced in South Africa. Against this background, certainly one of the most daunting challenges will be how to maintain standards and provide quality education to all learners in terms of their right to education (Bray, 2000).

According to Pandor, the challenge is not simply racial integration but the successful promotion of the values of dignity, equality, and the advancement of human rights and freedoms (Pandor, 2004, p. 11). In this regard the following immediate problems need to be addressed in South African education.

*Promoting Multilingualism*
From a macro perspective, the enrolment of learners appears to be deracialized to a large degree in most schools, except where there are impediments such as the language-medium of teaching and learning (Sujee, 2004, pp.43). According to Mda, language has been used as a basis for classifying and dividing people, and as the cornerstone of segregationist education policies (Mda, 2004, p. 11). Although the Language in Education Policy (Department of Education, 1997) intends to promote multilingualism and the development of all eleven official languages, there are many factors that still prohibit the realization of this policy. Many indigenous African

language speakers perceive English as offering greater socioeconomic and educational opportunities and as potentially "unifying" a linguistically diverse nation. Therefore, "students are very often not the authors of their own language; question/response cycles are predetermined and learnt by rote, and educators maintain a traditional controlling role by the use of questions where the answers are already known to and preconstructed by the educators themselves" (Ulrich, 2003, p. 1). An understanding of the interdependence of language and culture is central to all intercultural education situations. Advocates of intercultural theories emphasize the dialogical, interactive nature of learning, and accentuating the need to integrate the student experience and perspective as far as possible.

*Promoting Transforming Pedagogy*
Very often intercultural learning means perspective shifts or bigger transformations in one's way of understanding knowledge and the world. Knowledge creation has a cultural base, but traditional models of teaching marginally address only the impact of how educators and learners address this issue in the classroom. Our immediate concern is with the racial/cultural perceptions of educators. A key ingredient in building positive self-concepts in learners will be a culturally relevant educator. For educators to be culturally relevant in their teaching, they need to have positive images of their learners (Bekerman, 2003).

*Promoting Multiple Cultures*
It is imperative that educators have respect and concern for the intellectual welfare of individual learners that reveals cross-cultural differences (Humphrey, 2003). It is therefore critical that educators make a continuous effort to gain a working understanding of the cultural mores and practices of various ethnic groups (Dakwa, 2003). Interaction of learners in a classroom environment depends to a large part on the cultural training experience the individuals involved have received. Learners from cultures where there is considerable difference between the expectations for the genders will, for example, experience difficulties in classroom environments where male and female learners must work together on projects or assignments. Customs and practices that learners experience in an intercultural environment might clash with their personal value systems and produce distress.

*Promoting Diverse Religions, Beliefs, and Opinions*
The state has universally been accepted as the provider of education for its citizens. But many parents still think that the moral education of their children is solely both their right and duty. This is largely true of religious parents who are afraid that education in public schools could somehow harm their children. They appeal to their right to choose a specific (religious) education for their children that may, seemingly, result in a clash between the children and their parents when exercising the right to freedom of religion, belief, and opinion.

*Promoting Tolerance in Education*
In societies with citizens of diverse ethnic, racial, social, religious and cultural backgrounds, children must learn to socialize in order to fully participate in the society they live in (Erben and Dickenson, 2004). Keeping them isolated in schools within

homogenous communities may seriously limit their possibilities of choice and their chances for attaining their own good. The demands of tolerant behavior can be taught to children (Raulo, 2003), and as they become more rational, the reasons for promoting such behavior should also be given. However, the disposition of not condemning other people and their behavior will help to make tolerance less needed. Sometimes individuals try to appeal to the tolerance of others by staking a claim to cultural habits and the right to practice them. The idea behind cultural rights, however, is to protect only those features of different cultures that benefit their members in some way or other, not those features that harm them. The challenge in South Africa lies in creating a balance in promoting respect for the rights of cultural, religious, and linguistic communities (South Africa, 1996(a): s 185).

## Conclusion

The constitution has laid the foundation for peace, reconciliation, reconstruction, and justice in a new democratic and developing South African nation. It constitutes the bridge between the past, present, and the future; it provides the instruments or tools to guide people away from the past culture of authority to a culture of accountability—a culture in which every exercise of power should be justified. Nevertheless, transforming a nation and setting new values, norms, and principles is not an easy process: it is ongoing and cannot happen overnight. It involves complex constitutional and political issues but, more importantly, much deeper and fundamental social and cultural changes that impact on people's private lives and individual freedom.

Current strategies in South African school education address the ideal of equality of opportunity, equality of treatment, and equality of outcome. Creating equality in education by working together (Tirisano), introducing a national curriculum, and establishing democratic school governing bodies all imply a value-laden and value-based approach to education in South Africa.

Equality of treatment should promote the core human values of respect, compassion, just treatment, fairness, peace, truthfulness, and freedom (Nieuwenhuis, 2004, p. 55). Be that as it may, we face many challenges, and real and immediate problems that, in education, include securing safe schools and a peaceful education process; dealing with cultural diversity in the classroom and solving conflict caused by cultural differences; cultivating respect for diversity and differences; balancing equality of treatment with freedom to be different, to be oneself and the choice of one's schools; and ensuring quality education to all learners in terms of their right to education.

## Notes

1. Negotiations for peaceful change took place against the background of a collapsed apartheid system and increasing internal and international pressure. The negotiation process was not unique but a well-managed process. It used compromise-seeking and deadlock-breaking mechanisms to keep the negotiations on track; most decisions were taken based on consensus and compromise (De Villiers, 1994, pp. 1–11)
2. This process was essential because a culture of negotiation was established. It laid the foundations of democracy and planted the seeds for the birth of the constitution. A group of

leaders from all sectors of society kept the process on track and succeeded in building trust relationships between negotiators and their alliances, and this ultimately contributed to the success of the process.
3. The interim constitution (1993) also contained the constitutional principles upon which the 1996 constitution was later tested, evaluated, and approved.
4. Other basic characteristics of the constitution include: majority government, constitutional supremacy, a Bill of Rights, an independent judiciary, three spheres of government and the accommodation of diversity.
5. The constitution states, "all constitutional obligations must be performed diligently and without delay" (s 237). Similarly, the public administration, including the education administration, is bound by the provisions of the constitution and must deliver public services impartially, fairly, equitably, and without bias (s 195[1]). The public service must govern, inter alia, in terms of democratic principles, and on the basis of representivity, transparency, and accountability.
6. *Ubuntu* has been characterized as synonymous with humaneness, social justice, and fairness (*S v Makwanyane*, 1995, pp. 224–225, 237, 243, and 250; Rautenbach and Malherbe, 1999, pp.11–12).
7. For example, see the postscript of the interim constitution and the preamble of the constitution.
8. A National Peace Accord was established to pave the way for a new constitutional order in South Africa.
9. The TRC consisted of three committees: the Human Rights Violations Committee, the Rehabilitation, and Reparation Committee and the Amnesty Committee (Rakate, 2004, pp. 289–297).
10. Section 1 contains the founding values of the constitution.
11. For example, human dignity (s 10); equality (s 9); and freedoms such as freedom of religion, belief, and opinion (s 15); expression (s 16); and of association (s 18).
12. For example, the abolition of corporal punishment protects the learner's right to dignity, equality (of boys), privacy, security of the person, and freedom from cruel and inhuman treatment.
13. Section 28 must be read with subsection (2).
14. For example, for provision of basic nutrition, shelter, basic health care services and social services, and to protect the right to education: on school-going age (*Minister of Education v Harris*, 2001; on basic shelter and social services (*Government of the RSA v Grootboom*, 2000).
15. Section 9 deals with equality before the law and the right to equal protection and benefit of the law; affirmative action designed to advance and achieve the equal enjoyment by previously disadvantaged persons or groups of all their rights and freedoms; the prohibition of unfair discrimination by the state and other persons on grounds including race, gender, sex, pregnancy, ethnic or social origin, color, sexual orientation, age, disability, religion, culture and language; the adoption of national legislation to prevent or prohibit unfair discrimination.
16. Education legislation (for example, the South African Schools Act of 1996) and the individual provincial education laws give effect to human rights (including the right to equality) in the school environment.
17. The court held that "the guarantee of equality lies at the very heart of the Constitution. It permeates and defines the very ethos upon which the Constitution is premised" (*Fraser v Children's Court*, Pretoria North, 1997, p. 20).
18. Equality of outcome is also referred to as substantive equality (Albertyn and Kentridge, 1994, p. 149). For school education it may imply accelerated and enhanced education and training for previously disadvantaged learners and educators; making available more funding for previously disadvantaged schools, and so on.
19. The right to education (s 29) also prohibits discrimination based on race, for example, in the choice of language preferences at school and seeks to redress the results of past racially

discriminatory laws and practices in this regard (*Matukane v Laerskool Potgietersrus*, 1996); discrimination on the basis of race is also prohibited in the establishment and maintenance of independent (private) schools (subs (3)).
20. Only a few education (school) cases came before the constitutional court: for example, on human rights (*Christian Education South Africa v Minister of Education*, 2000).
21. Some of these institutions (for example, the Human Rights Commission) have played a leading role in promoting human rights and shaping the laws and policies of the country; the Commission for the Promotion and Protection of the Rights of Cultural, Religious and Linguistic Communities, on the other hand, has been struggling to get off its feet.
22. They guarantee the right to use the language and participate in the cultural life of one's choice (South Africa, 1996(a): s 30 and 31). Integrated schools changed from being homogeneous to multicultural schools.

## References

Albertyn, C. and Kentridge, J. (1994). Introducing the right to equality in the interim constitution. *South African Journal of Human Rights*, 10 (2): 149–178.
Asmal, K. (2001). *Implementation plan for Tirisano*. Pretoria: Department of Education.
*AZAPO v the president of the republic of South Africa* 1996 8 BCLR 1015.
Bekerman, Z. (2003). The influence of teachers' multicultural and bilingual perspectives on integrated Palestinian-Jewish education. In Lasonen, J. and Lestinen, L. (Eds.), *Conference proceedings, UNESCO conference on intercultural education* (7 pp.). Finland: University of Jyvaskyla, Institute for Educational Research.
Botha, H. (1994). The values and principles underlying the 1993 constitution. *South African Publiekreg/Public Law*, 9 (2): 233–244.
Bray, W. (2000). *Human rights in education*. Pretoria: Centre for Education Law and Policy.
Carl, A.E. (1994). "Education for peace"—Die rol van enkele opvoeders in Opvoeding vir Vrede. *South African Journal for Education*, 14 (1): 59–66.
*Christian Education South Africa v Minister of Education of the Government of the RSA* 1999 9 BCLR 951(CC).
Dakwa, K.D. (2003). Theme: Values, beliefs and controversial issues in pedagogy. Incorporating cultural awareness in the classroom: A socio-cultural approach. In Lasonen, J. and Lestinen, L. (Eds.), *Conference proceedings, UNESCO conference on intercultural education* (6 pp.). Finland: University of Jyvaskyla, Institute for Educational Research.
De Villiers, B. (1994). *The birth of a constitution*. Cape Town: Juta & Co., Ltd.
De Waal, J., Currie, I., and Erasmus, G. (2001). *The bill of rights handbook*. Cape Town: Juta.
Department of Education. (1997). *Language in education policy. Issued in terms of section 3(4)(m) of the National Education Policy Act 27 of 1996*. Pretoria: Department of Education.
Department of Education. (2000). *Revised Curriculum 2005*. Pretoria: Department of Education.
———. (2004). *Review of school governance in South African public schools*. Pretoria: Department of Education.
Ministry of education (2001). *Manifesto on values, education and democracy*. Pretoria: Department of Education.
Devenish, G.E. (1999). *A commentary on the South African Bill of Rights*. Durban: Butterworths.
Erben, M., and Dickinson, H. (2004). Basil Bernstein: Social divisions and cultural transmission. In Olssen (Ed.), *Culture and learning* (pp. 53–65). Connecticut: Information Age Publishing.
*Fraser v Children's Court, Pretoria North* 1997 2 BCLR 153(CC) .
*Government of the RSA v Grootboom* 2000 11 BCLR 1169(CC).

Humphrey, Ben. (2003). Values, beliefs and controversial issues in pedagogy. Cross-cultural diversity issues: Principles and processes to enrich international educators. In Lasonen, J. and Lestinen, L. (Eds.), *Conference proceedings, UNESCO conference on intercultural education* (pp. 1–7). Finland: University of Jyvaskyla, Institute for Educational Research.

*Matukane v Laerskool Potgietersrus* 1996 3 SA 223(T).

Mda, T. (2004). Education and multilingualism. In Nkomo, M., Chisholm, L., and McKinney, C. (Eds.), *Reflections on school integration* (pp. 163–182). Cape Town: HSRC Publishers.

Mureinik, E. (1994). A bridge to where? Introducing the interim Bill of Rights. *South African Journal on Human Rights*, 10 (1): 31–48.

*Minister of Education v Harris* 2001 11 BCLR 1157(CC).

Nieuwenhuis, J. (2004). From equality of opportunity to equality of treatment as a value-based concern in education. *Perspectives in Education*, 23(2): 55–64.

Pandor, N. (2004). Integration within the South African landscape: Are we making progress in our schools? In Nkomo, M., Chisholm, L., and McKinney, C. (Eds.), *Reflections on school integration* (pp. 11–18). Cape Town: HSRC Publishers.

Rakate, P.T.K. (2004). *The duty to prosecute and the status of amnesties granted for gross and systematic human rights violations in international law: Towards a balanced approach model.* Unpublished LLD thesis. Pretoria: University of South Africa.

Raulo, M. (2003). Academic professionalism. Multicultural education—Education in tolerance? In Lasonen, J. and Lestinen, L. (Eds), *Conference proceedings, UNESCO conference on intercultural education* (8 pp.). Finland: University of Jyvaskyla, Institute for Educational Research.

Rautenbach, I.G., and Malherbe, E.F.J. (1999, 3rd ed.). *Constitutional law*. Durban: Butterworths.

*S v Makwanyane* 1995 6 BCLR 665(CC).

Soudien, C., Carrim, N., and Sayed, Y. (2004). School inclusion and exclusion in South Africa: Some theoretical and methodological considerations. In Nkomo, M., Chisholm, L., and McKinney, C. (Eds.), *Reflections on school integration* (pp. 19–42). Cape Town: HSRC Publishers.

Sujee, M. (2004). Deracialisation of Gauteng schools—A quantitative analysis. In Nkomo, M., Chisholm, L., and McKinney, C. (Eds.), *Reflections on school integration* (pp. 43–60). Cape Town: HSRC Publishers.

South Africa (RSA). (1996a). *Constitution of the Republic of South Africa of 1996*. Cape Town: Government Printer.

South Africa (RSA). (1996b). *South African schools act 84 of 1995*. Cape Town: Government Printer.

Ulrich, N. (2003). Our culture is like normal: The student's perception on learning about foreign cultures. In Lasonen, J. and Lestinen, L. (Eds.), *Conference proceedings, UNESCO conference on intercultural education* (6 pp.). Finland: University of Jyvaskyla, Institute for Educational Research.

# Section II
## Teachers and Students

## Chapter Five
## Color Coded: How Well Do Students of Different Race Groups Interact in South African Schools?

*Saloshna Vandeyar and Heidi Esakov*

### Introduction and Study Context

Since 1994, various policies have been unveiled and legislation enacted to hasten the process of desegregation in the schooling system of South Africa. Policymakers have been extremely adept at formulating policy in terms of the desegregation of schools. School management teams in a desperate bid of survival have taken these policies and filtered it down to the mesolevel for implementation purposes. Teachers at best have tried to adapt their teaching strategies to the changing schooling scenario. The focus has been on policymakers and school governing body members and the intricacies and subtleties that surround the relationship between politics and policies. Yet, the most important variable in the teaching-learning triad has been overlooked, the student. It would seem as if policymakers assumed that by placing students from diverse cultural backgrounds within close proximity of each other, all will bode well; that the legacy of segregation will be forgotten and that students will naturally just mix and get along with each other and adapt to the hegemonic school culture.

"The apartheid education system of South Africa entrenched gross educational disparities and inequities between different racial groups" (Sayed 2001, p. 252). The education and training system under apartheid was characterized by three key features: First, the system was fragmented along racial and ethnic lines, and had been saturated with the racial ideology and educational doctrines of apartheid. Schools were fragmented into 19 different education departments and funding varied on the basis of "race." Second, there was a lack of access or unequal access to education and training at all levels of the system. Vast disparities existed between the provisions of education for black and white. Third, there was a lack of democratic control within the education and training system.

This fragmented, unequal, and undemocratic nature of the education and training system had profound effects on the development of the economy and society.

It resulted in the destruction, distortion, and neglect of the human potential of South Africa with devastating consequences for social and economic development.

The need for rectification and parity in all aspects of education was thus a necessary imperative in a new, democratic education system (Sayed, 2001; Soudien, 1994). The Bill of Rights and the South African constitution set the process of school reform in motion. A number of fundamental and social human rights are stipulated in these documents, such as the right to education, redressing past discriminatory practices, and language in education (Section 29 of the Bill of Rights). Besides the constitution, the South African Schools Act (Act no. 37 of 1997) formalized the desegregation of schools in South Africa, and created the opportunity for students from diverse cultural backgrounds to be together. It was hoped that in creating this opportunity students would become integrated into the whole school environment and the seed of a new society will be sown.

These policies set the stage for desegregation to unfold at schools, by establishing the physical proximity of members of different groups in the same school; however, it did not go further to interrogate the quality of contact—not only in the personal attitudes of students and educators but also in the institutional arrangements, policies, and ethos of the school (Carrim, 1995; Sayed, 2001). Accordingly, this study asks, what indeed is the nature of the student relationships and interactions within and outside the school, and to what extent has the school organized itself to provide positive relations among black and white students?

To assist in addressing the research problem, the following two research questions were formed to guide the study: (1) To what extent has the ethos of these schools been transformed toward integration in the truest sense and how do the students perceive this in practice? (2) Are new forms of self-identity beginning to emerge?

The argument is presented as follows. We begin by outlining from current research a theoretical framework on the reinvention and renegotiation of the construct "race." We then describe the sample and context and the research methodology that was implemented. This is followed by the development of themes that emerged from interviews. We conclude with an analysis and discussion of findings with respect to the quality of contact between students and examine ways in which schools have elected or omitted to adopt certain strategies with the "opening" of racially exclusive schools in South Africa.

### Reinventing and Renegotiating the Construct of "Race"

Interrogating the concept of race has been and still is a problematic transnational discourse (McCarthy and Crichlow, 1993; McCarthy et al., 2003; Nieto, 2000; Winant, 2000). During the early years of the twentieth century, Du Bois and Boas proposed a revolutionary interpretation of race, debunking the traditional theory of race as an essential biological "truth" (Winant, 2000). Rather, race as an inconstant sociohistorical construct dictated by economic variables was proposed and is now the commonly accepted view (Carrim, 1995; Dolby, 2001, 2000; McCarthy and Crichlow, 1993; McCarthy et al., 2003; Morrison, 1993; Nieto, 2000; Winant, 2000).

The mercurial nature of race and racial analysis is a discernible signifier of the social interactions through which perceptions of whiteness and blackness are formulated;

whiteness as the subject, blackness as the object or "the other" (Dolby, 2000, 2001; Hall, 1996a, 1996b; Morrison, 1993). Reliant on each other for their interpretations (McCarthy et al., 2003), whiteness and blackness are constantly shifting connotations of the political and social context in which they are situated (Dolby 2001, 2002; McCarthy et al., 2003; Morrison 1993). However, the simplistic polarization of these constructs has served to entrench views of homogeneity rather than dispel them, and in its wake has left race and racial experience being recast as static and essentialized (Hall, 1996a, 1996b). These reductionist conceptions of race, casting whiteness as the norm against blackness, the subaltern, have, unofficially, sanctioned the continuation of racism (Dolby, 2000, 2001; Nieto, 2000; Morrison, 1993). In consequence, with institutions such as schools reflecting this erratic social discourse, differential standards of societal power, privilege, and positioning are consigned to whiteness (Delpit, 1988), relegating blackness to inconsequentiality:

> society categorizes people according to both visible and invisible traits, uses such classifications to deduce fixed behavioral and mental traits, and then applies policies and practices that benefit others. (Nieto, 2000, p. 35)

The polarities of whiteness and blackness can further be embodied as first and third world constructs respectively. However, the contestation between the first and third world, whiteness and blackness (Dolby, 2002), has been problematized by the advent of globalization (McCarthy et al., 2003). With the accompanying escalation of cultural and human transborder migration, race has taken on a new identity (McCarthy et al., 2003). Identities are increasingly hybridized and influenced by cross-racial popular culture. In this blurring of cultural and economic borders, race is constantly being reinvented and renegotiated through mediums such as popular culture (Dolby, 2000; Wasserman and Jacobs, 2003).

## Sample and Context

Following a combination of purposive and convenience sampling, we identified and interviewed 29 Grade 8 students at 3 secondary schools in Pretoria, South Africa.

For convenience the schools will be referred to as Broadstream, Silversands, and Ridgewood. In each school, it was popularly assumed that rapid desegregation had been implemented since 1997. Information available suggested that the race profile of the teaching cadre had remained relatively unaltered. It was thus suspected that there would exist considerable mismatches between the linguistic and cultural backgrounds of teachers and a significant proportion of students at each school. What was not clear was how desegregation at the level of classroom practice was manifesting itself. Of particular interest were interactions between students. Table 5.1 presents a summary of the school profiles.

## Research Methodology

This study aimed at capturing the nature of interactions among Grade 8 students in schools and to investigate Grade 8 student perceptions of interactions and their feelings

**Table 5.1** Profile of Schools

|  | Broadwater | Silverstream | Riverwood |
|---|---|---|---|
| Type of school | Ex-Model C[1], well-resourced school | Ex-Model C | Ex-Model C |
| Location | Middle- to upper class predominantly white suburb | Middle- to upper class predominantly white suburb | Middle-class white suburb |
| Medium of instruction | English | English | English |
| Established | 1902 | Early 1976 | 1963 |
| Pre-1994—catered exclusively to | White English-speaking students | White Afrikaans-speaking[2] students | White students, the majority of whom spoke English |
| Post-1994—student population | 1375 (55% white, 34% African, 8% Indian, and 3% colored)[3] | 970 (55% African, 33% white, 7% Indian, and 5% colored) | 1080 (41% African, 42% white, 12% Indian, and 5% colored) |
| Total number of grade 8 students | 280 | 227 | 252 |
| Teachers | 81 (73 white [37 GDE[4]; 36 SGB[5]]; 4 African [2 GDE; 2 SGB]; 3 Indian SGB; 1 colored SGB | 46 (37 white [24 GDE; 13 SGB]; 2 Indian SGB, 2 Colored GDE and 5 African [3 GDE; 2 SGB] | 59 (56 white [32 GDE; 24 SGB], 1 Indian SGB, 1 Colored SGB, and 1 African SGB |
| Teacher—student ratio | 1:28 | 1:28 | 1:25 |
| School fees | R 8,300–00 | R 7020–00 | R 6250–00 |
| Mode of transportation | Privately owned cars | Taxis | "Bussing-in"[6] phenomenon |

of a sense of belonging to the school. The study was both descriptive and investigative in nature. A qualitative-interpretive methodology was undertaken. The data gathering techniques that were used in this study included *interviews, observation,* and *field notes*. It would not be an exaggeration to say that interviews were the main data gathering technique used in this study and were "semistructured" in nature.

A purposive sampling method with an emphasis on difference in race was used to identify students. Approximately twelve students were selected from each school. The criterion for selection was to have an approximate number of students in terms of gender, across the four race groups: African, white, colored, and Indian. Selected students were required to obtain parental consent to participate in this study. Of the 35 students selected, 6 refused to participate in the study. Hence, a total of 29 students were interviewed. Questions comprised of five broad categories and were open-ended. All interviews were recorded. In order to get a better feel of the schooling and learning environment, various field notes were made, based on informal observations of these schools. Informal conversations were conducted with some teachers. Attention was also given to the physical mien of the school, which included observations of artifacts such as paintings, décor, photographs, portraits, and school magazines. Data gathered was coded according to themes generated during the study and from relevant literature reviewed in the field.

## Findings of This Study

Three major findings emanated from this study.

### The "white way" is the "one way"

Contrary to legislation and the "spirit of reconciliation" and despite considerable changes in the racial composition of these schools, it does not appear that these schools have shown much flexibility in accommodating the multitude of cultures and worldviews that black students have bought along with them. Jay, an Indian male from Ridgewood, aptly sums up the general attitude: "Because like at school, there's like, you can say the whites, like only believe in like one way, like you do one thing like one way." As appears to be the trend in the vast majority of desegregated schools in South Africa, students are expected to adjust their outlooks and their identities, molding themselves to the cast of the dominant culture. Consequently, this study found that assimilation still dominates the schools' approach (Carrim, 1998; Vally and Dalamba, 1999). Essentially, students are expected to act and behave "white." Students wanting to "fit in" are placed under tremendous pressure to conform and be absorbed into the system.

The institutionalized racism of these schools, which is so obviously attempting to maintain the status quo, is apparent in the composition of teachers. At Ridgewood, for instance, compared to a multitude of white teachers, there was one colored, one African, and one Indian—a *veritable* reflection of the "rainbow nation"![7]

A problematic phenomenon that has been observed at desegregated schools with a predominantly white teaching staff is that teachers tend to "teach to" white students, and ignore black students in the class (Nieto, 2000). Aware that they are being ignored the students are inclined to attempt to gain their teachers' attention by being disruptive and rebellious (Nieto, 2000).

Many black students at these schools were not proficient in the language of instruction. They were thus doubly disadvantaged, as they had to cope with processing abstract concepts through an unfamiliar language, as well as having to accomplish this without the support of their teachers. What also often tends to be the case is that the intelligence of these students is judged by their ability to express themselves in English. The result is what Robert Merton referred to as "the self-fulfilling prophecy," whereby students perform according to the expectations of their teacher (Nieto, 2000, p. 43). In accordance with this term, students view their own ability as lesser than their peers who are fluent in English, and as a result underperform.

### "We are equal and we care for one another" versus "They say we are stupid": Student Commitment versus Institutional Racism

In spite of the seeming indifference of the majority of teachers in initiating the dismantling of racial divides, students appear to be instinctively working toward unification.

Most students expressed the fundamental necessity of confronting past injustices so as to tackle the present social inequities in order to "give us another point of view from where people come from," (Thabo, African male, Silversands), and achieve true national reconciliation. This finding significantly diverges from the Vally and Dalamba report's (1999) negative appraisal of student commitment and serves as an inspirational signifier of South Africa's future. It is quite extraordinary how students have, on their own accord, embarked on the process of deconstructing the artificial boundaries imposed by the formed myth of race as a given truth, and are striving toward integration. The finding of this research is a momentous step forward and can only be done justice through the testimonies of students:

> Yes, because some people . . . must learn to understand that there is no difference between us, just color difference. So they must understand that we can be together. (Jason, white male, Ridgewood)
>
> I just think that we should come to terms that we are equal and that we care for one another. (Sipho, African male, Silversands)
>
> [to] understand that we are not racist and you know, we can be friends. (Costa, white male, Silversands)

Teachers were not actively promoting integration and at one school in particular they were reinforcing boundaries. When we asked Sipho how the different races mixed in his class, he answered: "There are no whites and no Indians, just Africans." According to students there are two homogenous classes: one African and one virtually all white; the other classes are fairly evenly mixed. This unacceptable nuance of segregation is not only inconsistent with the moral obligation schools have in providing equitable education, but contravenes legislation. It emerged that the "white" class comprised the academic achievers, as Michelle a white female from that class told us: "In my class this year we've only got like four African people, the rest are white and then there are two Indians. I'm in the best Grade 8 class and the people in there are really clever." Conversely, Sipho told us that the students in his class are rebellious and are viewed by the teachers as failures. On probing as to how teachers responded to their rebellion, Sipho replied: "they don't like it, they say we are stupid." Taken aback by such a revelation, we went on to ask him how this made him feel. Sipho, casting a downward glance replied: "[we] feel down . . . like we're nothing."

## "I sound and talk like a black person": Emergence of New Self-Identities

There was the virtual absence of race in the students' "self-articulation" (Carrim, 2000) of their identities. On a positive note, this could be interpreted as the refusal of the youth of South Africa to conform to the racial categories of old that were fundamental instruments in the abuse of human rights. Yet, on a more realistic level, students are aware of the societal stigma attached to openly talking about race (Makhanya, 2004, p. 19). Natalie (white female, Broadstream) succinctly summed up the prevailing attitude: "I don't know, people feel uncomfortable talking about other races in front of that race." In an attempt to discard the label of oppressor, many white students adopted a color-blind approach. As Candice, a white female from Broadstream insisted, "I never see color."

Research has indicated that what has often been the consequence is that black students have had their sense of identity "whitened" in order to have a feeling of belonging. Essentially, borrowing Berger's (1980) term of "the male gaze," they have come to view themselves through the eyes of the hegemonic order; "the race gaze." A resulting factor of the "race gaze" is that students are torn between two seemingly contradictory identities. Mpkazi (African, female, Broadstream) highlighted this issue: "I'm not there with it [identity/culture] if you can understand. I'm just a bit different." She narrated an incident with another student who asked her: "why don't you speak with a kind of African accent like other African people?" Mpkazi mentioned how this comment had troubled her as, "she may mean that I'm not in touch with my roots. I'm not African enough. She might mean that I act too white or something." Straddling two worldviews has left Mpkazi interpreting her identity through the eyes of another, in turn becoming the other.

What this study has come to show is that, unlike Mpkazi, many students across racial lines are beginning to question and reject this static and forged identity. Rather, within a country that now celebrates difference, there is an explicit move toward the reclamation of the self. It appears that some students are beginning to dabble in th construction of a personal identity, using the medium of universal popular cultures to express their shifting selves (Wasserman and Jacobs, 2003). Ironically, it is often these students who are labeled as dissenters for not conforming: "[teachers] don't like us because we don't do what we supposed to do . . . they don't understand me, because they don't know who I am they think that I'm somebody who I'm not" (Lee-Anne, white female, Ridgewood). This student's nonconformist and laid-back attitude, inspired, by Reggae, has been a causative factor in her being labeled rebellious by her teachers.

### Analysis and Discussion of Findings

The "white way" is the "one way"

The assimilatory approach implemented by these schools not only attempts to alter a student's identity but also, by implication, sends a subliminal message to the student that her culture is subaltern (Nieto, 2000). With identity and value systems being so intrinsically linked, this process contributes toward the strengthening of the lifechances of the protectorates of the hegemonic culture, by eroding away at the "devalued" student's self-esteem. The following comment by Mpkazi encapsulates this trend of identity denigration and the need to conform to the dominant culture: "It's kind of like you are wondering why and maybe it's because we are a bit slow in progressing into . . . like their level." The prevailing ethos of "whiteness" at Broadstream has not only caused Mpkazi to question her own worth as embedded in her "background," but has also made it quite clear to her, that in order for her to find her place in the school, she must be absorbed into the dominant culture: "progress[ed] into their level." However, as this research shows, many students across racial lines are beginning to question and challenge this subjugation of identities.

The implications of a basically white teaching staff cannot be underestimated. Filtering through from the white-macroculture's belief of their supposed superiority is the polarized belief of black ineptitude (Jansen, 2004). This belief is played out in the allied microculture of the school, sending an unambiguous message to black students of not only the incompetence of black teachers, but by implication, their own. Black students are unable to fully identify with their white teachers. Consequently, endeavoring to conform to the supposed norms that the teacher as an agent of the culture of power projects, the student's identity is further denied. As a result, this largely homogenous teaching fraternity serves to further entrench the process of assimilation and deprive students of their cultural and linguistic heritage.

The tendency to marginalize black students became particularly evident during interviews with students from Ridgewood. Interestingly it was a white female student, Lee-Anne, who critically exposed the prevalence of this practice of racial inequity at her school: "Well some of the teachers are fine with different colors and the 'differentness' of us, but others exclude them. They don't understand why black students are doing what they doing, or for what reason . . . so black students get a bad reputation . . . they are in the class but teachers are always against them and don't pay any attention to them . . . and black students just rebel against it."

In effect, teachers are silencing the academic voices of black students. This "de-voicing" of the students can also be seen as an attempt by white teachers to diminish the self-worth of black students. This is indicative of a covert form of institutional racism, which serves to restrict the access of nonwhites to "power and privilege" (Sleeter, 1993).

This silencing figuratively extends beyond the walls of the classroom and permeates the power structures that have been conferred to students. Power translates into a prefect body, which is a select group of students in Grade 12 who preside over other students, implementing the school rules. In a sense they are the foot soldiers whose role is to further enforce the ethos of the school. Their role is particularly influential as they also serve as the mouthpiece of students. Students are fully aware that their success at the school, that is the possibility of being selected as a prefect and having their voices heard, are intrinsically linked to and dependent on their association with the "ruling" culture (Delpit, 1988). Subsequently the "whiteness" of a student's attitude is a significant factor in the selection of the prefects. In questioning why the school has never had an African head prefect, Mpkazi subliminally yields to the positioning as the less significant "other," as projected through the eyes of the authoritative voice of the school: "[I] think that there are more white prefects because we're aren't really equal or something."

Racism does not always refer to the words or actions resulting in the subjugation of another, but often that which is not said and done to challenge existing oppression and prejudice, so as to maintain and ensure a position of privilege and power (Nieto, 2000). Most forms of structural racism that we came across were cloaked and at times, even subliminal. Institutionally, there appears to be very little active attempts to catalyze the process of integration.

"We are equal and we care for one another" versus "They say we are stupid":
Student Commitment versus Institutional Racism

The situation is far from ideal, yet six years after the publication of the Vally and Dalamba report (1999) there appears to be a distinctive move by students toward a synthesizing of previous cultural and racial blocs. A definite trend by students of taking the initiative, where the institution has not, to promote cross-cultural and racial relationships was observed. According to most students, on the whole, there appears to be noteworthy racial mixing in the classrooms. As Emily, a white female from Broadstream affirms, "it's basically mixed but different people react different ways. There are many that are totally mixed, there are a couple of groups though, about two groups in the class, this is the Afrikaner side and this is the African side, its just exactly like that, they talk to each other, there is no borders or anything really."

At all these schools, students, for the most part, grouped together in cultural and racial pockets during breaks. Twelve years after apartheid has formally been abolished, it appears that racial divides are still a prevalent aspect of the psyche of South Africans (Lund, 2004; Moodley and Adam, 2004). One student reasoned: "[students interact with their race and cultural groups at break] because they have the same culture, the same way of thinking, I don't know, they feel secure in people who are like them" (Emily, white female, Broadstream). Yet, it was Natalie's pragmatic remark that signified and accentuated the repercussions of this racial and cultural pocketing: "Some students do not like to mix with students of a different race but, they don't tell you I'm not going to hang out with you because you black, they just move away from you . . . in time they are so far away that you are not going to talk to them anymore."

The effects of apartheid and the consequences of the regime's propaganda have permeated every aspect of the South African psyche. Apartheid's objective went beyond the ideology of racial segregation; the myth of white intellectual supremacy was fabricated in an attempt to justify the immorality of this abuse of human rights. A damaging consequence of this has been the ingrained perception of the inferior intellectual aptitude of Africans (Lund, 2004). As Sipho's peers can testify, a decade of democracy has not seen the debunking of this myth in all strata of society. This inexcusable derision of the students' intellectual capability and sense of worth is nothing short of abuse. In all probability the impact of this conduct is far-reaching and the scars will be indelibly imprinted on the students' sense of self.

Although this research has uncovered disturbing incidents of race-based discrimination, many teachers are genuinely committed to the progress of all their students. Teachers have in many ways been abandoned by the system. It has been unrealistic to believe that teachers will have adopted the skills to cope in a multiracial and cultural society, one that is often alien to their own life-experiences. Currently there are a limited number of preservice and inservice teacher training programs (Moletsane, Hemson, and Muthukrishna, 2004). Although these programs are offered in the form of an additive to the mainstream curriculum they would undoubtedly give teachers not only coping mechanisms, but may have the potential of opening up their worldview, and creating a classroom environment where social justice and equity prevails.

Looking to the schools for guidance would prove futile for teachers. Schools have implemented an additive form of multiculturalism, which merely glosses over the basic aspects of major religions and cultures represented in South Africa. Emily affirmed, "We never go onto lines that are religion based, cultural based, what has happened in the past, anything like that. We just stay on the syllabus really. You know there is nothing, no wondering or discussions or anything like that." As expressed and criticized by students, there is a notable absence of the interrogation of power structures that underlie societal relations. Rather, the form of multiculturalism practiced appears to be an attempt to contain the multiplicity of worldviews and value systems and is divorced from the reality of societal influences.

Apart from the political transformation in South African schools over the past 11 years, there has also been a dramatic move toward self-expression amongst students who have began questioning that, which was previously accepted as truth. This reclamation of the individual voice has left many teachers, who were shaped by the apartheid system, struggling to adjust to the inquiring nature of current students. Instead of facing the challenge of political, social, and personal analysis, teachers have opted to maintain safe ground. However, students, untainted by apartheid's ideology of insular acceptance, have voiced their frustration of having to accept the comfort zone of ignorance retained by their teachers. Realizing the consequence of ignorance, students are thirsting for the confrontation of contentious issues that are obstacles to equity:

> People need to know, they need to know what happened, they need to get some feelings out. You know have a discussion, not let it get too heated, but actually have a good discussion about it. It would help. (Emily)

*"I sound and talk like a black person": Emergence of new self-identities*
Catalyzed by globalization, division that characterized the South African identity years ago has given way to a hybridized identity (Wasserman and Jacobs, 2003). The emergence of new identities and cultural self-perceptions must be interpreted in the transglobal flow of cultural and economic capital (McCarthy et al., 2003). Emerging from decades of isolation from the international community, South Africa's current role as a player in the global village has resulted in psychological and cultural borders becoming ever more indistinct. Previously defined boundaries have become blurred and the rigid racial categories of old are being subverted and replaced by fluid interpretations of what it means to be a South African (Wasserman and Jacobs, 2003). Jethro's articulation of his identity captures this inconstant hybridism: "I'm colored, I have Indian family, and I have a hip-hop culture."

Research into popular culture in the South African schooling context has made known the conjunctive association of specific races with a particular form of popular culture (Dolby, 2000). It was shown that popular culture served to redefine notions of whiteness through fictional links with a conceptualized view of eurocentricity, or what Dolby refers to as "global whiteness" (Dolby, 2000). An aspect of this popular culture of global whiteness took the form of techno music (Dolby, 2000). Played against this new imaging of whiteness was an abstract blackness, conveyed through the popular culture of hip-hop, rap and RMB. HIP-hop, rap, and rmb find their origins in black African American music. Some white learners were identifying with this type of music as part of popular culture and did not associate it with blackness.

However, what has emerged from this study is that a blurring of these illusory borders and synthesizing of cultures is taking place on the school grounds (Tshoagong, 2004). Jason, a white male, from Ridgewood, is emblematic of these rearticulated forms of identity. Intrigued by a white student using rap and hip-hop-style jargon, we asked Jason what style of music he likes and whether he felt that it is in any way representative of his identity: ". . . hip-hop, Rap House music, RMB . . . people like get a style and a thing from music . . . yes, you could actually quite say that [that this music is representative of the way he perceives his identity]." We asked Jason whether he felt that his teachers have an understanding of who he is as a person. To this he rejoined, "No, some of the teachers don't because I hang out with a lot of African people during break and so the teachers say, the white teacher Mr. P says that I sound and talk like a black person, actually I don't care, that's me who I am . . . if he doesn't like the way I talk then he must just leave me alone."

## Conclusion

Conceivably, the greatest problem with identity formation in South Africa is apartheid's legacy of projection of identities, instead of the self-composition (Sonn, 1994): "whiteness" (the thesis) as a construct fabricated through "blackness" (the antithesis) to ensure privilege and power (Dolby, 2000; Morrison, 1993; Sonn, 1994). It is more than a decade since democracy, yet not all South Africans seem to have dismantled the archaic racial classifications of the psyche, and this prefabricated manner of viewing "the other" has permeated the current mindset of many teachers and schools.

However, South Africa is no longer an isolated country resilient to transglobal changes. Consequently, globalization and its cultural offshoot of popular culture are currently informing and influencing the ever-shifting identities of South Africa's youth. In this study it emerged that young South Africans are subliminally rejecting racial categories of old. This emergence of new self-identities, refusing to be contained by prescriptions of the past, has seen students like Jason, Lee-Anne, and Jethro becoming the synthesis, discarding homogeneous identities for an unfixed and inclusive "South African" identity.

Agony and ecstasy are universal and it is not only South Africans who can learn from our mistakes of past and present. To move forward we must acknowledge our stumbling blocks—as well as our strengths. South Africa is rich in human potential, and the value of our diversity and the fluidity of identities should be seen and used as an asset. Although this study has found that attitudes of schools and teachers have resisted change, it has become apparent that students are constructing new transracial and cultural identities and actively working toward a South Africa where "the spirit of *Ubuntu*—that profound African sense that we are human only through the humanity of other human beings—is not a parochial phenomenon, but [adds] globally to our common search for a better world" (Cryws-Williams 1997, p. 82).

## Notes

1. Model C school—a government attempt to cut state costs by shifting some of the financing and control of white schools to parents.
2. Afrikaans is one of the 11 official languages recognized by South Africa's new constitution. In the previous dispensation, only English and Afrikaans were recognized as official languages and languages of instruction in white, Indian, and colored schools.

3. The terms colored, white, Indian, and African derive from the apartheid racial classifications of the different peoples of South Africa. The use of these terms, although problematic, has continued through the postapartheid era in the country. In this chapter, we use these terms grudgingly to help present the necessary context for our work.
4. GDE—Provincial Department of Education.
5. SGB—School Governing Body.
6. "Bussing-in"—A phenomenon that has occurred post-1994, where large numbers of African students are transported by bus from neighboring black suburbs to middle-class English medium schools.
7. Rainbow nation—reflective of the diverse racial, cultural, and language groups of people in South Africa. Limited English Proficiency—the learner comes from a home in which a language other than English is primarily used for communication and who has difficulty in understanding, speaking, reading, or writing the English language.

## References

Berger, J. (1980). *Ways of seeing*. Harmondsworth: Penguin Books Limited.
Carrim, N. (1995). From "race" to ethnicity: Shifts in the educational discourses of South Africa and Britain in the 1990s. *Compare*, 25 (1): 17–35.
———. (1998). Anti-racism and the "new" South African educational order. *Cambridge Journal of Education*, 28 (3): 301–320.
———. (2000). Critical anti-racism and problems in self-articulated forms of identities [1]. *Race Ethnicity and Education*, 3 (1): 25–44.
Delpit, L. (1988). Power and pedagogy in educating other people's children. *The Harvard Educational Review*, 58: 280–298.
Dolby, N. (2000). Changing selves: Multicultural education and the challenge of new identities. *Teachers College Record*, 102 (5): 898–911.
———. (2001). White fright—The politics of white youth in South Africa. *British Journal of Sociology of Education*, 22 (1): 5–17.
———. (2002). Making white: Constructing race in a South African high school. *Curriculum Inquiry*, 32: 7–29.
Hall, S. (1996a). New ethnicities. In D. Morley and C. Kuan-Hsing (Eds.), *Stuart Hall: Critical dialogues in cultural studies* (pp. 441–449). London: Routledge.
———. (1996b). What is "black" in black popular culture? In D. Morley and C. Kuan-Hsing (Eds.), *Stuart Hall: Critical dialogues in cultural studies* (pp. 465–475). London: Routledge.
Jansen, J. (2004). *Race, education and democracy after ten years—How far have we come?* Prepared for the Institute for Democracy in South Africa (IDASA), Lessons from the Field: A Decade of Democracy in South Africa.
Lund, T. (2004, January). Colour blind. *Fairlady* (pp. 22–26).
Makhanya, M. (2004, September 5). Whites must come on board. *Sunday Times* (p. 19).
Cryws-Williams, J. (Ed.), (1997). *In the words of Nelson Mandela: A little pocketbook*. Parktown: Penguin Books.
McCarthy, C. and Crichlow, W. (1993). Introduction. In C. McCarthy and W. Crichlow (Eds.), *Race identity and representation in education* (pp. xiii–xxix). New York: Routledge.
McCarthy, C., Giardina, M.D., Harewood, S.J., and Park, J. (2003). Contesting culture: Identity and curriculum dilemmas in the age of globalization, post colonialism, and multiplicity. *Harvard Educational Review*, 73: 449–465.
Moletsane, R., Hemson, C., and Muthukrishna, A. (2004). Educating South African teachers for the challenge of school integration: Towards a teaching and research agenda. In M. Nkomo, C. McKinney, and L. Chisholm (Eds.), *Reflections on school integration: Colloquium proceedings* (pp. 61–79). Cape Town: HSRC Publishers.

Moodley, K.A. and Adam, H. (2004). Citizenship education and political literacy in South Africa. In J. Banks (Ed.), *Diversity and citizenship education: Global perspectives* (pp. 159–183). San Francisco: Jossey-Bass.
Morrison, T. (1993). *Playing in the dark—Whiteness and the literary imagination.* London: Picador.
Nieto, S. (2000, 3rd ed.). *Affirming diversity: The socio-political context of multicultural education.* New York: Addison Wesley Longman, Inc.
Sayed, Y. (2001). Post-apartheid educational transformation: Policy concerns and approaches. In Y. Sayed and J. Jansen (Eds.), *Implementing education policies: The South African experience* (pp. 250–271). Cape Town: UCT Press.
Sleeter, C. (1993). How white teachers construct race. In C. McCarthy and W. Crichlow (Eds.), *Race identity and representation in education* (pp. 157–171). New York: Routledge.
Sonn, J. (1994). Establishing an inclusive, democratic society: The need for a multicultural perspective in education. *Multicultural Teaching,* 12 (3): 9–13.
Soudien, C. (1994). Equality and equity in South Africa: Multicultural education and change. *Equity & Excellence,* 27 (3): 55–60.
Tshoagong, D. (2004, August 28). Wiggaz: Whites with black souls (and style). *Saturday Star,* p. 7.
Vally, S. and Dalamba, Y. (1999). *Racism, "racial integration" and desegregation in South African public secondary schools: A report on a study by the South African Human Rights Commission* (SAHRC). Johannesburg: SAHRC.
Wasserman, H. and Jacobs, S. (2003). Introduction. In H. Wasserman and S. Jacobs (Eds.), *Shifting selves: Post-apartheid essays on mass media, culture and identity* (pp. 15–28). Cape Town: Kwela Books.
Winant, H. (2000). The theoretical status of the concept of race. In L. Back and J. Solomos (Eds.), *Theories of race and racism—A reader* (pp. 181–190). London: Routledge.

# CHAPTER SIX
# CHALLENGES IN INTEGRATED EDUCATION IN NORTHERN IRELAND

*Claire McGlynn*

## Introduction

In a country emerging from protracted ethnic conflict, efforts to reconcile divided factions through the integration of education appear both admirable and logical. In Northern Ireland integrated education has been officially established since the first planned integrated (mixed Catholic, Protestant, and other) postprimary school opened in 1981. Despite the opening since of a further 56 primary and postprimary schools, the majority set up by groups of committed parents, education in Northern Ireland remains largely segregated with children either attending Catholic-maintained schools or de facto Protestant-controlled schools. A small proportion also attend a growing Irish medium sector and a very small number of independent schools. The number of children attending integrated schools is less than 6 percent and the phenomenon has been described as voluntary integration by parental consent rather than compulsory desegregation (Gallagher and Smith, 2002).

This chapter considers the contribution of integrated schools and looks to their future in Northern Ireland, in a context of emerging government policy, demographic decline, funding pressures, and curriculum change. Drawing on a series of interviews with principals and other key stakeholders, it explores issues such as the perceived contribution of integrated education to peace building, the significance of variation in practice between schools, increasing pupil diversity, teacher education, and the lessons for other postconflict societies.

## Does Integrated Education Work?

A variety of theoretical approaches have been applied to integrated education in Northern Ireland including intergroup contact (Allport, 1954) to promote intergroup acceptance (Hewstone, 1996; Pettigrew and Tropp, 2000). Social identity theory (Tajfel, 1978) and self-categorization theory (Turner, 1991) have also been employed to investigate the relationship between identity and conflict, with social identification as "Catholic" and "Protestant" being a defining feature of the conflict. In Northern Ireland intergroup boundaries are usually impermeable (Breen, 2000; Breen and Hayes, 1996). There is limited evidence (McGlynn, 2001) suggesting that two of the integrated schools in Northern Ireland encourage decategorization and recategorization in the long term. A further perspective that has been applied to integrated education (McGlynn, 2003) is that of critical multicultural theory (Kincheloe and Steinberg, 1997; Mahalingham and McCarthy, 2000; Nieto, 2000; Sleeter, 2000) that proposes that a celebration of diversity divorced from a serious questioning of social inequality may be fraudulent and potentially harmful. Ideas about the management of diversity in the integrated schools in Northern Ireland are particularly pertinent as the concept of the melting pot (Glazer and Moynihan, 1963) concedes to that of the salad bowl (Esteve, 1992) wherein the flavors of different cultural communities remain distinctive.

Due to methodological, logistical, and ethical difficulties, research into the impact of integrated education in Northern Ireland has been limited. However, a picture is emerging of a form of education capable of impacting positively on identity, outgroup attitudes, and forgiveness and reconciliation, with the potential to help rebuild the social cohesion fragmented by protracted conflict (McGlynn et al., 2004). Findings reported to date include an increase in the number of intercommunity friendships (Irwin, 1991; McClenahan, 1995; McGlynn, 2001) amongst those attending or having attended integrated schools. The work of Stringer, Wilson, Irwing, Giles, McClenahan, and Curtis (2000) suggests that such schools influence social attitudes through intergroup contact, finding that pupils of mixed or integrated schools assume an integrative approach to key issues such as marriage and education.

In a first study of the impact of integrated education on two cohorts of past pupils, McGlynn (2001) found a significant long term positive impact on cross-community friendships, respect for diversity, confidence in plural settings, and an enhancement of the ability to empathize with alternative perspectives. Integrated education also appeared to facilitate student exploration of personal and group identities in a nonthreatening environment with a subsequent range of impacts on past pupils' perceptions of their social, religious. and political identities. A superordinate "integrated identity" was claimed by the majority of these former students and was characterized by respect for diversity, broadmindedness, understanding, and tolerance.

A further study of undergraduate university students by Niens, Cairns, Hewstone, and McLernon (2003) found that students with experiences of integrated education were more inclined toward forgiveness. Carter (2004) recognizes the criticality of the sustained positive contact between mixed cohorts in Northern Ireland's integrated schools, contact that can nurture both relationships and collaborative learning. However, direct contact itself may not necessarily be a prerequisite for community relations benefits. Paolini, Hewstone, Cairns, and Voci (2004) propose that in Northern Ireland just knowing that a person has friendship with a member of the Protestant or Catholic outgroup can reduce levels of sectarian prejudice via an anxiety-reducing mechanism. Thus whilst the integrated schools currently account for less than 6 percent of the school population their impact may, through a ripple effect, in fact be significantly wider.

## The Current Climate

Whilst evidence for the benefits of integrated education slowly accumulates, it is important to gain an insight into the current sociopolitical and economic climate in which it is developing. It is a truism that peace comes slowly after protracted conflict. Although considerable time has elapsed since the first IRA ceasefire of 1994, progress toward a more peaceful and democratic society has been painstaking, characterized by a lack of trust on all sides, a struggle to reinstate local government rule and the continued violent oppression of communities by paramilitary organizations. Indeed some commentators have observed that the Belfast/Good Friday Agreement (Government of the United Kingdom of Great Britain and Northern Ireland, the Government of Ireland, 1998) has served only to institutionalize sectarianism resulting in greater political and social segregation. In such a climate, the critical absence of a long-term vision for Northern Ireland that might guide and inform educational reform has been highlighted (McGlynn et al., 2004).

The long awaited Policy and Strategic Framework for Good Relations in Northern Ireland "A Shared Future" has the potential to provide such a vision, aiming over time to establish "a normal, civic society in which all individuals are considered as equals" (Office of the First Minister and Deputy First Minister [OFMDFM], 2005, p. 8). With a key challenge of new policy being the building of cohesive communities, Shared Future critically states that "separate but equal is not an option," exposing the unsustainability of parallel services (OFMDFM, 2005, p. 20). This would appear to have clear implications for an education system segregated along denominational lines, not the least in the current climate of economic cutbacks and considerable demographic decline. Indeed an economic imperative might drive educational change more forcefully than a social one. Proposed sweeping reforms of public administration also have the potential to transform educational services.

Into this mix, one must also add the replacement of academic selection by parental choice in a much contested reform of the mechanism by which children proceed from primary to postprimary education; a greater expectation of resource sharing between schools (Department for Education Northern Ireland [DENI], 2004); and also significant reform of the national curriculum characterized by a reduction in content, an increase in pupil choice, an emphasis on transferable skills and values, the introduction of Personal Development and Local and Global Citizenship as statutory areas, and a return to greater teacher autonomy (Council for the Curriculum, Examinations and Assessment [CCEA] 2002, 2003). This, then, is the complex and changing climate in which integrated education must compete.

A demand for more integrated education in Northern Ireland is evident. Under the 1989 Education Reform (Northern Ireland) Order the government has a duty to meet the needs of parents requesting it where it is feasible. In September 2004, over 670 applicants to integrated schools were turned away due to the lack of places, and surveys of the Northern Ireland population reveal that 82 percent support integrated education, 72 percent would choose an integrated school if it were close and had comparable academic standards, and 81 percent consider it important to peace and reconciliation (Millward Brown on behalf of Northern Ireland Council for Integrated Education [NICIE], 2003). The evidence suggests that the reason that parents do not send their children to integrated schools is simply insufficient places (Gallagher and Smith, 2002). As demand continues to outstrip supply (McGlynn, 2004a; Morgan and Fraser, 1999; O'Connor, 2002), the Integrated Education Fund [IEF] pledges financial support to parents and newly established schools. To this end it enjoys high profile support from both the business and international community.

An equally compelling argument for integrated education comes not just from parents, but from young people themselves, a voice that has regrettably long been absent from decisions regarding education in Northern Ireland. A number of recent studies demonstrate the needs of young people. Based on talkshops with 194 16–17-year-olds across Northern Ireland, Ewart, and Schubotz (2004) report that the most important request was to improve community relations for young people by providing more formally integrated schools. A further study of primary and postprimary children by Kilpatrick and Leitch (2004) reveals the pupils' view that sustained and long-term contact is the key to the success of cross-community initiatives. In addition the pupils stated a clear preference for encounter and discussion of views with young people from different cultural backgrounds.

## Methodology

The objective of this study was to gain an understanding of the perspectives of school principals and other key educational leaders with regard to the contribution of integrated education to society in Northern Ireland and how the sector might develop in the future, their current priorities and concerns, their thoughts on variation in values and practice between integrated schools, and insight into the lessons for other countries moving out of conflict.

The sample consisted of six principals selected from a sample of more than 10 percent of the 57 integrated schools, namely 3 primary (one Grant Maintained [GMI] more than 10 years old, 1 GMI less than 5 years old and 1 transformed Controlled Integrated [CI]) and 3 equivalent postprimary schools. The sample also included urban and rural variation. In addition a purposive sample (Punch, 1998) of key individuals were interviewed, that is, a director of NICIE, a director and a senior officer from the IEF, 2 high-ranking officers from the DENI, and a high ranking officer from the OFMDFM, making a total of 12 individuals.

Intensive semistructured interviews of between one and two hours were used to allow for flexibility, exploration at greater depth, and to facilitate the emergence of rich data. Units of relevant meanings were clustered and common themes determined before themes general and unique to all interviews were identified with any significant individual differences noted (Cohen and Manion, 1994).

## The Contribution of Integrated Education to Peace: Different Constructs

The principals, although managing schools of varying type, size, location, and stage of development were united in their conviction that the shared daily experience of learning in an integrated school breaks down barriers, develops friendships, and broadens the mind of the children in their care. Furthermore they saw their schools as impacting positively on the wider family circle, including parents and grandparents, as the latter were challenged and brought together by school activities. Some felt that some parents had deliberately chosen a form of schooling for their children that was different from their own. Whilst the officers of the government departments were understandably keen to promote the role of all schools in developing a more peaceful society, they perceived that a strength of the integrated schools was the active embracing of diversity. In particular OFMDFM agreed that good relations goals were at the core of integrated education.

Could it be, however, that there is a mismatch between the expectations of school and society? Whilst integrated education has always had vociferous opponents and cynics (O'Connor, 2002) it can be tempting to reduce the debate to an overly simplistic "integrated good, segregated bad" formula that denies the efforts of the nonintegrated sector. NICIE admits that the language of the integrated movement has not always been helpful in this respect and that may have led to a social construct of integrated education as occupying the moral high ground and communicated a sense of it being the only way to do peace education in schools, leading to unrealistic expectations of its potential.

The IEF officer dealing with national and international publicity is adamant that too much is expected, placing a heavy burden on integrated schools to single-handedly solve the problems of Northern Ireland. When members of the local community say to the principal of the new postprimary integrated school in this study "Good luck—you are just what this place needs," how realistic are their expectations of what it might achieve? The principals were all secure in their construct that their schools were merely part of a solution and could not be a panacea. One suggested that their indicators of success may be different from those of society, namely the demonstration only of academic successes. All of the principals were convinced that parents demand a good academic education for their children, with the social benefits sometimes coming as a secondary or even irrelevant by-product. As such there may be significantly different constructs of integrated education subscribed to by school leaders, parents, government, and also wider society. Expectations and indicators of success and failure will thus vary accordingly, indicating a clear need for a coherent brand that realistically outlines what integrated education means in practice and what it can offer to society in Northern Ireland.

## Variation in Practice and the Need for a Coherent Concept

In the meantime could it be that there are as many varieties of integration as there are integrated schools? Montgomery, Fraser, McGlynn, Smith, and Gallagher (2001) identified three general ways in which integrated schools approached the concept of integration, namely passive (allowing it to happen naturally), reactive (doing something if the need arises), and proactive (planned policy and structures). In a study of a third of the primary integrated schools, Loughrey, Kidd, and Carlin (2003) suggest that community relations practices are more substantial and sophisticated in GMI than in CI schools.

One reason for such variation might be the location and context of the school and it is clear from the six principals in this study that the schools are situated in areas of greatly varying degrees of local opposition/support and sectarian tension. It is also apparent that the schools are at different stages of developing their practice with regard to integration and that this might be reflected in the type of integration observed. For example, the CI primary school has a history of mixed pupil intake going back over a century whilst the new GMI postprimary is less than a year old. For the CI postprimary school Catholic pupils are also nothing new and in addition it traditionally caters to a number of service children from a local army base, for whom attendance at a Catholic school would be unlikely. Common to all the schools is a higher than average number of children with special educational needs. All of these factors are likely to influence the practice of integration. Whilst development toward proactive practice might occur at a different rate depending on the school context; what is important is that the principals recognized that progress on that journey is imperative.

Questioned on variation in practice a number of GMI principals explained that integration could be unintentionally squeezed out by an examination-led system and by demands such as securing permanent accommodation and recruiting new staff. The four GMI principals stated that they constantly reminded their staff and pupils of the importance of integration in practice suggesting that integration can exist as a theme in parallel with academic priorities. One primary GMI principal acknowledged the varying degrees of commitment to integration that she had observed at meetings with other principals but countered that whilst some CI schools did little, others were more proactive than some GMI schools. One CI principal expressed frustration at being constrained by his board of governors who thought that he was embracing integration too quickly.

In schools that have been integrated for longer there may also be an issue around a "normalization" effect identified by principals who perceive this as a mark of a school's success where the new norm becomes integration. The external segregated world becomes abnormal and issues of who is Catholic and who is Protestant become largely irrelevant. This was qualified, however, by two GMI principals who spoke of the need to avoid complacency and continually advance understanding of what integration means in practice. The two CI principals were in agreement with NICIE that constant awareness of difference is an artificial place to be and in general they appeared more reluctant to place an overt emphasis on difference than the GMI principals. It remains to be seen if approaches that emphasize either categorization or decategorization will actually result in different outcomes but considering the importance of social identification in Northern Ireland, a critical discussion of such strategies is important.

Does current variation in practice imply that integrated education is a flawed product and that those who promote it, including the All Party Parliamentary Group (APPG) who recently launched their Integrated Education Manifesto (IEF, 2004), are misleading the public? Possibly so, although such judgment depends on the readers' construct of the sector. Endorsement seems to spring from a conviction that it is good to educate children together in a postconflict society, without perhaps a deep understanding of the real processes involved and the implications of managing diversity differently. Constructive criticism of passive schools is healthy but it becomes increasingly clear that there is a overarching need to create a coherent understanding of what integration as opposed to coexistence means, to communicate clearly this to society, to identify and share good practice and to find ways of encouraging and supporting all schools in their development of it, whatever their starting point.

A joined up concept of integrated education could be informed by a shared framework of values. NICIE proposed that values should be evidenced by the quality of relationships and interaction and be measurable by an inspector. Three concepts were identified by the principals, NICIE and the IEF as being central to integrated education, namely equality, interdependence, and inclusivity. The latter was articulated both in the sense of an emphasis on all kinds of difference—children with special needs, those of other faith, those who do not believe in any religion, and children from ethnic minority backgrounds. Whilst principals were resistant to schools being mirror images of each other they were keen to engage in a debate around what integration is and what distinguishes an integrated school. Of paramount importance here is the need to distinguish between integrationist and assimilationist practice. The degree of school self-evaluation of these practices varied greatly.

One way of coordinating a shared framework of values could be through the Statement of Principles of integrated education (NICIE, 2005), agreed in 1991 and subsequently subscribed to by all integrated schools. In the six schools there was a varying awareness of these, although the principals were conscious of contentious NICIE proposals to reduce parental representation on governing bodies. Two GMI principals indicated that the most important aspect was the practice of integrated principles but another described the written version as "antique furniture," nice to show people but rarely used. There was some support for debate leading to a revision of those principles, in particular to include the viewpoints of schools, including the CI schools, set up after 1991.

## Increasing Pupil Diversity and Teacher Education

Another change has been the increasingly visibility of ethnic and religious minority groups in Northern Ireland, and McGlynn (2003) has exhorted schools to develop practice in line with critical multicultural theory. Despite the size and rural location of some of the integrated schools in this study, all had a number of pupils from ethnic minority backgrounds and there was evidence of developmental work beyond the national curriculum, including projects on world faiths and on the theme of racism, in all six schools. The necessity of responding thoughtfully to the pupil intake, in the spirit of the integrated ethos, was apparent and this was seen to enrich the learning experiences of pupils and staff alike. For example, the new GMI postprimary principal described ways in which his school celebrated the first languages spoken not only by pupils but also by migrant staff who do the cleaning. It would appear that the schools are making attempts to recognize increasing diversity and should be encouraged to explore how this might fit with existing theory and practice regarding the management of difference.

Multicultural practice is, however, new and challenging and one GMI primary principal pointed to the lessons learned in her school. During one lesson taken by a member of the Muslim community, she had been obliged to gently intervene when it became apparent that the young children thought they had to go to Mecca. An increase in the number of children other than Catholic or Protestant also raises issues regarding admissions criteria and the IEF recommends that DENI modify these to allow a redefinition of all minority children as nonmajority, giving a fairer chance to families from nontraditional backgrounds.

In all six schools teacher education for diversity and integration was identified as an ongoing area for development. A variety of in-service practices exist ranging from simply learning on the job in the CI primary to a more sophisticated package of in-house and external support in the longer established GMI postprimary. Complexity of teacher education did not, however, necessarily map on to the age or type of school. This appeared to be dictated more by a state of readiness to address integration as identified by the school principal. Less than a year after his school opened, the principal of the new GMI postprimary declared his staff sufficiently mature to begin to unravel their personal baggage in order to become better integrated. The principal of the CI postprimary also felt that his staff were now primed for such work.

One principal saw teacher education as part of the journey toward integration, suggesting that the process itself is more important than a final destination. What training then is needed for this journey? The answer is likely to be highly complex for not only are the teachers at different stages with regard to knowledge and understanding of integration, but so too are the nonteaching staff, the governors, and the parents. Whilst the OFMDFM points to a need to grapple with this in initial teacher education in Northern Ireland (Hagan and McGlynn, 2004), the ongoing teacher education for integration currently falls mainly on the schools. The principals then have the unenviable task of meeting those diverse needs and guiding the whole school toward a coherent model of integration.

With external funding NICIE itself has embarked on a number of projects to begin to address this deficit. A recent manual (Lynagh and Potter, 2005) provides practical advice and resources to help all schools develop an inclusive learning community and in addition to its induction course for teachers new to integrated education, NICIE now also offers a "Learning with Diversity" program to teachers from both within and outside the integrated sector. Demand, however, exceeds NICIE's ability to supply. Thus there is considerable scope for institutions of continuing professional development to work together with NICIE and the Education and Library Boards to support schools in their journeys toward inclusion and integration.

## Lessons for Other Countries Moving Out of Conflict

The appropriate training of teachers is only one of a number of issues highlighted by the leaders of integrated education in Northern Ireland. Whilst Gallagher (2004) is clear that no one particular type of educational response can promise better intercommunity relations around the world, other societies might heed the messages sent from those involved in integrated education in Northern Ireland. All those interviewed were united on one fundamental point—transforming a society that has suffered from ethnic conflict is a slow process and outcomes cannot be expected too soon. Where to start is also problematic but ultimately this may hinge on assessing readiness in parts of society to listen to each other and to begin to work in partnership. A key message is that integrated education should not be imposed, rather hearts and minds need to be won over to the belief that learning together is beneficial. One

GMI principal exhorts other societies to consider integrated education because there will always be people brave enough to embrace it whilst others caution that there will inevitably be resistance to change. Others speak of the need to make the process less of a struggle for parents and families. In Northern Ireland, integrated education is almost exclusively parent-driven and there are clear calls to the government to incentivize it.

NICIE also points out that the balance of representation and responsibility between ethnic groupings throughout the whole school is crucial as is the effective delivery of an antibias curriculum that does not shy away from controversial issues, with neither working in isolation. It is also considered essential to begin integrated education with children in the early years and to provide a continuum to age 18. It was pointed out that integrated education brings, more than anything else, children together, and to this end the development of the school as a shared community space is an important aim. There is an apparent willingness amongst the integrated education movement to share its mistakes as well as its successes.

## The Future of Integrated Education in Northern Ireland

What does the future hold for integrated education in Northern Ireland? There is parental support but as one GMI principal points out it is the practice and not the aspiration that counts. Certainly it is equitable to expect the government to provide integrated education to those requesting it (as per the 1989 Order), but DENI is anxious that this is qualified by economic feasibility.

Representatives of the sector and the government officials interviewed recognize that the current climate of demographic downturn and economic cutbacks make the transformation of existing schools to integrated status a better option than the establishment of further new schools. Previous research has challenged the credibility of integrated education in transforming schools that have patently not transformed and who display little appetite for the prospect (O'Connor, 2002). The IEF, however, calls for an increased awareness in the transformation option to encourage those who had not previously considered it.

A further solution might be to extend existing integrated schools, and indeed additional places have already been provided by this means. The difficulty with this option is that it could have a detrimental impact on the balance of school populations in some areas. For the first time there are signs of a degree of competition for pupils between integrated schools. DENI and OFMDFM were optimistic of a new educational environment requiring greater sharing of estate and resources in order to deliver pupil entitlement and choice, predicting a considerable growth in interschool collaborations. However, the integrated principals in this study were generally skeptical about the practicality of such arrangements.

Whilst some differences were observed in discussions between CI and GMI principals, most notably a lessened emphasis on difference in the former, all the principals were convinced of the participation of their school in a journey toward greater integration. McGonigle, Smith, and Gallagher's (2003) study of six transforming schools stressed the importance of developing this shared understanding. In this study it seemed that the GMI schools had a better understanding of what such a journey might entail. It is clear that the integrated sector would benefit greatly from establishing an understanding of what integration means, to identify and disseminate best practice and to develop mechanisms for supporting schools in their respective journeys. A more coherent brand for integrated education would enable indicators of success to emerge and could lead to more realistic expectations from wider society.

It is increasingly clear that integrated schools are not the only ones doing work in the field of community relations. Recent initiatives signpost an increasing willingness of other parts of the education system to be seen to be embracing diversity and its challenges in schools. The introduction of citizenship into the national curriculum also requires all schools to address issues of a divided and diverse society. Whilst welcoming this and freely admitting to "not being the only show in town," NICIE does, however, claim to be the only one with a balanced staff, pupil, and governing body structure. Despite the current context the IEF is confident of reaching its target of 10 percent of children in integrated schools by 2010, although it predicts that prevailing factors will cause other schools to become more mixed. The OFMDFM admits that after change in postprimary admissions and economic rationalization, an ultimate reconfiguration of education in Northern Ireland will occur, but questions the extent to which this should be planned. Those in the integrated sector would argue that it is precisely this type of strategic planning for integration that should take place. DENI cannot be seen to favor one type of schooling over another and exercises its obligations under the 1989 Order by providing funding for NICIE and for integrated schools. Most of the principals are critical of what they see as the lack of a strategic government plan regarding integrated education. NICIE interprets this as the risks associated with starting new schools being absorbed by NICIE rather than by the government. In the current climate a pragmatic solution might be for DENI to promote an integrated option where rationalization is needed but even this is contentious and OFMDFM is adamant that local solutions must be negotiated.

Some of the principals called for an enhanced role or statutory function for NICIE that would enable it to coordinate a more coherent approach across the integrated schools but DENI has no plans for this at present. Frustrated by current DENI restrictions regarding its role, NICIE, in conjunction with the IEF, has recently secured external funding to develop a lobbying function and to research, identify, and disseminate good integration practice, including schools and practitioners from outside the integrated sector.

In a fragile postconflict Northern Ireland it has been claimed that whilst the rights of religious communities to their own form of schooling are undeniable, such schools may not rebuild social cohesion as readily as integrated schools where contact between children is sustained (McGlynn, 2004b). The integrated schools have certainly challenged the appropriateness of church involvement in the management of schools in divided societies (Smith, 2001). If the integrated education movement is to realize its potential contribution to a still divided society it is clear that a number of issues must be addressed. First, there is a need for the sector and DENI to open up a debate as to the meaning and objectives of integration (as opposed to coexistence) in education, including an investigation of strategies for the management of difference in integrated schools that might support the long-term reconstruction of Northern Irish society. Second, such objectives need to be clearly communicated throughout the education system and society, leading to a coherent and recognizable "integrated" brand with independently measurable indicators of success. Third, practical and sustained support must be planned and provided for those teachers and schools brave enough to embark on the challenging journey toward integration.

## References

Allport, G.W. (1954). *The nature of prejudice*. London: Addison-Wesley.
Breen, R. (2000). Class inequality and social mobility in Northren Ireland 1973 to 1996. *American Social Review*, 65: 392–406.
Breen, R. and Hayes, B. (1996). Religious mobility in the United Kingdom. *Journal of the Royal Statistical Society, Series A*, 159 (3): 493–504.
Carter, C. (2004). Education for peace in Northern Ireland and USA. *Theory and Research in Social Education*, 32 (1): 24–38.
Cohen, L. and Manion, L. (1994). *Research methods in education*. London: Routledge.
Council for the Curriculum, Examinations and Assessment [CCEA]. (2002). *Detailed proposals for the revised primary curriculum and its assessment arrangements*. Belfast: CCEA.
———. (2003). *Pathways—proposals for curriculum and assessment at key stage three*. Belfast: CCEA.
Department of Education for Northern Ireland [DENI]. (2004). Future post-primary arrangements in Northern Ireland: Advice from the post-primary review working group. Bangor: DENI.
Education Reform (Northern Ireland) Order (1989). S.I 1989, No. 2406 (NI20) Belfast: Her Majesty's Stationery Office.
Esteve, J.M. (1992). Multicultural education in Spain: The autonomous communities face the challenge of European unity. *Educational Review*, 44 (3): 255–272.
Ewart, S. and Schubotz, D. (2004). *Voices behind the statistics: Young people's views of sectarianism in Northern Ireland*. London: National Children's Bureau.
Gallagher, T. (2004). *Education in divided societies*. Hampshire and New York: Palgrave Macmillan.
Gallagher, T. and Smith, A. (2002). Selection, integration and diversity in Northern Ireland. In A.M. Gray, K. Lloyd, P. Devine, G. Robinson, and D. Heenan (Eds.), *Social attitudes in Northern Ireland: The eighth report*, pp. 120–137. London: Pluto.
Glazer, N. and Moynihan, D.P. (1963). *Beyond the melting pot: The Negroes, Puerto Ricans, Jews, Italians and Irish of New York City*. Cambridge, Massachusetts, and London: MIT Press.
Government of the United Kingdom of Great Britain and Northern Ireland, the Government of Ireland. (1998). *The Agreement; agreement reached in the multi-party negotiations*. Belfast: Northern Ireland Office.
Hagan, M. and McGlynn, C. (2004). Moving barriers: Promoting learning for diversity in initial teacher education. *Intercultural Education*, 15 (3): 243–252.
Hewstone, M. (1996). Contact and categorisation: Social psychological interventions to change intregroup relations. In C.N. Macrae, C. Stangor, and M. Hewstone (Eds.), *Foundations of stereotypes and stereotyping* (pp. 323–368). New York: Guilford.
Integrated Education Fund [IEF]. (2004). All Party Parliamentary Group [APPG]: Integrated Education Manifesto. Belfast: IEF.
Irwin, C. (1991). *Education and the development of social integration in divided societies*. Belfast: Department of Social Anthropology, Queen's University.
Kilpatrick, R. and Leitch, R. (2004). *Teachers' and pupils' educational experiences and school based responses to the conflict in Northern Ireland*. Belfast: Save the Children.
Kincheloe, J.L. and Steinberg, S.R. (1997). *Changing multiculturalism*. Philadelphia, PA: Open University Press.
Loughrey, D., Kidd, S., and Carlin, J. (2003). Integrated primary schools and community relations in Northern Ireland. *The Irish Journal of Education*, 34: 30–46.

Lynagh, N. and Potter, M. (2005). *Joined-up: Developing good relations in the school community*. Belfast: The Corrymeela Community and NICIE.

Mahalingham, R. and McCarthy, C. (2000). *Multicultural curriculum: New directions for social theory, practice and policy*. New York: Routledge.

McClenahan, C. (1995). *The impact and nature of intergroup contact in planned integrated and desegregated schools in Northern Ireland*. University of Ulster at Coleraine. PhD thesis.

McGlynn, C. (2001). *The impact of post primary integrated education in Northern Ireland on past pupils: A study*. Belfast: University of Ulster at Jordanstown. Unpublished EdD thesis.

———. (2003). Integrated education in Northern Ireland in the context of critical multiculturalism. *Irish Educational Studies*, 22 (3): 11–28.

———. (2004a). Education for peace in integrated schools: a priority for Northern Ireland? *Child Care in Practice*, 10 (2): 85–94.

———. (2004b, April). *Diversity of school provision in Northern Ireland—Co-existence of integration?* A paper presented to the Annual Conference of the Educational Studies Association of Ireland, Maynooth.

McGlynn, C., Niens, U., Cairns, E., and Hewstone, M. (2004). Moving out of conflict: The contribution of integrated schools in Northern Ireland to identity, attitudes, forgiveness and reconciliation. *Journal of Peace Education*, 1 (2): 147–163.

McGonigle, J., Smith, A., and Gallagher, T. (2003). *Integrated education in Northern Ireland: The challenge of transformation*. Coleraine: UNESCO Centre, University of Ulster.

Millward, B. (2003). *Public opinion survey: Integrated education in Northern Ireland*. Belfast: Northern Ireland Council for Integrated Education.

Montgomery, A., Fraser, G., McGlynn, C., Smith, A., and Gallagher, T. (2003). *Integrated education in Northern Ireland: Integration in practice*. Coleraine: UNESCO Centre, University of Ulster.

Morgan, V. and Fraser, G. (1999). When does good news become bad news? Relationships between government and the integrated schools in Northern Ireland. *British Journal of Educational Studies*, 47 (4): 364–379.

Niens, U., Cairns, E., Hewstone, M., and McLernon, F. (2003, September). *Intergroup contact in education: Impact on forgiveness*. Paper presented at the Conference for Peacebuilding after Peace Accords, Joan. B. Kroc Institute for International Peace Studies, University of Notre Dame.

Nieto, S. (2000). *Affirming diversity: The sociopolitical context of multicultural education*. New York: Longman.

Northern Ireland Council for Integrated Education [NICIE]. (2005). Latest News. Retrieved July 7, 2005 from www.nicie.org.uk.

O'Connor, F. (2002). *A shared childhood: The story of the integrated schools in Northern Ireland*. Belfast: Blackstaff Press.

Office of the First Minister and Deputy First Minister [OFMDFM]. (2005). *A shared future: Policy and strategic framework for good relations in Northern Ireland*. Belfast: OFMDFM.

Paolini, S., Hewstone, M., Cairns, E., and Voci, A. (2004). Effects of direct and indirect cross-group friendships on judgements of Catholics and Protestants in Northern Ireland: The mediating role of an anxiety-reduction mechanism. *Personality and Social Psychology Bulletin*, 30: 770–786.

Pettigrew, T.F. and Tropp, L.R. (2000). Does intergroup contact reduce prejudice? Recent metanalytic findings. In S. Oskamp (Ed.), *Reducing prejudice and discrimination* (pp. 93–114). Mahwah, NJ: Earlbaum.

Punch, K.F. (1998). *Introduction to social research: Quantitative and qualitative approaches*. London: Sage Publications.

Sleeter, C. (2000). *Critical multiculturalism and curriculum analysis*. New Orleans: AERA 2000 Annual Conference.

Smith, A. (2001). Religious segregation and the emergence of integrated schools in Northern Ireland. *Oxford Review of Education*, 27 (4): 559–575.

Stringer, M., Wilson, W., Irwing, P., Giles, M., McClenahan, C., and Curtis, L. (2000) *The impact of schooling on the social attitudes of children.* Belfast: The Integrated Education Fund.

Tajfel, H. (1978). *Differentiation between social groups: Studies in the social psychology of intergroup relations.* London: Academic Press.

Turner, J. (1991). *Social influence.* Milton Keynes: Open University Press.

# Chapter Seven
# Developing Palestinian-Jewish Bilingual Integrated Education in Israel: Opportunities and Challenges for Peace Education in Conflict Societies

*Zvi Bekerman*

### Introduction
### The Political, Social, Cultural, and Educational Background of Israel's Bilingual Initiative

*The Political*

As much as any other modern nation-state, the State of Israel is a product of invented tradition (Hobsbawm, 1983) and has institutionalized itself by establishing public education, a standardized legal system, and a secular equivalent to the church (Ben-Amos and Bet-El, 1999; Gellner, 1997; Handelman, 1990).

The Palestinian-Jewish conflict has an unfortunately long history starting with the birth of political Zionism at the end of the nineteenth century to the development of Arab nationalism in response to colonialization in the Ottoman and the British Empires in the nineteenth and twentieth centuries (Abdo and Yuval-Davis, 1995). Since the 1920s, violence in the region has intensified, particularly as a result of the UN partition decision in 1947 and when Prime Minister David Ben-Gurion declared the independence of Israel in 1948 without declaring state borders.

The 1948 war, called the War of Independence by the Israelis and the Nakba (the Catastrophe) by the Palestinian, was the first open military clash between the Zionism and Arab nationalist movements. Four major wars have subsequently erupted: The Suez Crisis of 1956 with Egypt; the Six-Day War of 1967 with Syria, Egypt, and Jordan; the Yom Kippur War of 1973 with Egypt and Syria; and the war in Lebanon of 1982. In 1977 and 1993, a peace agreement was signed between Israel and Egypt.

Since 1967, the eye of the storm has centered around the Israeli military and the Palestinian population in the 1967 conquered territories. The Palestinians in these areas, most of whom are children of Palestinian refugees expelled from Israel during the war of 1948, comprise today a population whose exact number remains contested (Khalidi, 1994; Kimmerling and Migdal, 1993). According to some, in the year 2000 this population approximated four million. The Intifada outbreaks in 1997 and 2000 organized under the flag of the Palestinian Liberation Organization (PLO), an organization formed in the late 1960s by Yassar Arafat, brought about even bloodier events that have yet to find a peaceful solution even after the Oslo agreement reached between the Israeli government and the PLO in 1993.

Currently the Palestinian population within the internationally recognized borders of Israel is 20 percent while the Jewish population is about 78 percent (CBS, 2001). Aside from the approximate 10 percent of Palestinian Israelis living in mixed cities (in mostly segregated neighborhoods) the majority live in segregated villages or small cities (Ghanem, 2001a).

*The Social*
From its inception, Israel—as stated in its Declaration of Independence—has been committed to full political and social equality for all its citizens irrespective of religion or ethnic affiliation. However, even the Israeli government agrees that Israel has not been fully successful in implementing this ideal, predominately implementing segregationist policies toward its non-Jewish minorities—policies that only recently are starting to be challenged in the courts of justice (Gavison, 2000).

These separatist policies were often ad hoc arrangements, products of military emergencies that accompanied Israel's development over the past hundred years. Though outcomes have varied, the most visible segregation can be seen in residential and educational arrangements that are fully separated for both the Palestinian and Jewish communities (Rouhana, 1997).

The Palestinian-Israeli conflict has strongly influenced Palestinian-Jewish relations that in turn have profoundly influenced the formation of Palestinian identity within Israel's borders. The Palestinian presence in the State of Israel and the awakening of Palestinian national consciousness has problematized the seemingly natural construct of the Israeli nation. The Jewish-Palestinian conflict remains the most explosive conflict in Israel, placing the Jewish majority and the Palestinian (primarily Muslim) minority at perpetual odds.

Palestinian identity has gone through various stages of development, tightly connected to their sense of citizenship and Israel's position in international politics. Following 1948, Palestinians—perceived as a threat by Jewish Israel—suffered from military governance including restrictions on their civic rights. All municipal, educational, social, and religious institutions were supervised to prevent the emergence of Palestinian centers of power (Ghanem, 2001). Such supervision, though somewhat relaxed, continues today. Since 1967, and with the incremental abolishment of military governance, major changes have taken place. Today Palestinian Israelis have gathered enough strength to challenge Israeli Jewish hegemony (Ghanen, 2001) and to change their self-identity from Israeli-Arab to Palestinian (Rouhana, 1998; Suleiman, 2004 ). These changes were supported by the uprising of their brethren in the conquered territories. In spite of structural differences that limit their civic rights and position them as second-class citizens because of unequal access to economic, political, and social resources, most Palestinians in Israel say they would rather stay in Israel than move to a Palestinian state if one were established (Smooha, 2004). Palestinians experience Israel as a Jewish "ethnic" state and not a democracy (Ghanem, 1998; Rouhana, 1998). From their perspective, Israel is a colonizing power that took their lands (Stasiulis and Yuval-Davis, 1995).

Both peoples have been plagued with tragedy and suffering. To paraphrase Edward Said (1994), we are dealing with two asymmetrical communities with symmetrical fears. While structurally the communities reflect a sharp asymmetry, their perceptions of the situation are similar. Both sides believe they have a monopoly on the objective truth of the conflict and on the identification of the perpetuating villain. These perceptions undermine prospects for conflict resolution (Bar-Tal, 1990, 1998).

Abu-Nimer (Abu-Nimer, 1999) identifies three main factors predicating the larger Palestinian-Jewish conflict in Israel. First, since 1948 the Palestinian people have lacked a political framework for self-determination. Second, Israel being defined as a Jewish State results in socioeconomic and political inequalities. Finally, juxtapositions emerge between the traditional and nonindustrial social and cultural structure of Palestinian society and Israel's Western-type social and cultural structure.

*The Cultural*
In their historical development and cultural resources, Jews and Palestinians do not represent dichotomous groups. However, given the long history of conflict and identities shaped in direct opposition to one another, they have been constructed as such (Kelman, 1999; Sharoni and Abu-Nimer, 2000). Religiously, both groups belong to the monotheistic tradition, and Moslem as well as Christian Palestinians see their roots in the Jewish prophetic tradition. Both groups include individuals representing a wide gamut of religious practices. Palestinians are composed of a majority of Moslems. Others include Christians (10 percent), Druse, and Bedouin (Barakat, 1993; Central Bureau of Statistics, 1999). Jews include groups ranging from ultraorthodox to fully secular with roots in both western and middle eastern countries.

Both Hebrew and Arabic, as predominant languages in Israel, share Semitic origins. However, the policy of the Zionist movement encouraged Jewish immigrants to adopt Hebrew, thus promoting language as a major source of identity and a cultural boundary marker (Ben-Rafael et al., 1994). The Palestinian Arabic dialect serves similar purposes for the Palestinian population in Israel. The Israeli population, whether Jewish or Palestinian, shares Hebrew as the primary language of communication. Jews have a very low Arabic literacy level with the exception of those coming from east Mediterranean countries.

Given the Ashkenazi (Jews arriving from western countries) hegemonic dominance in Israel, cultural differences beyond language can be perceived. Jews tend to emphasize individualism, low-context speech, monochronic time, and lower power distance where as Arabs tend to emphasize collectivism, high-context speech, polychronic time, and high power distance (Feghali, 1997; Hofstede, 1980) although reifying these differences is dangerous since considerable variability exists within each group and members of both groups are comfortable with both sets of cultural patterns. It is possible to conceptualize these differences based not on ethnic but educational and socioeconomic level.

*The Educational*
The many sociopolitical conflicts that beset Israel are reflected in the country's separate educational sectors that include nonreligious Jews, religious national Jews, orthodox Jews, Druze, and Arabs (the preferred name for the largest non-Jewish Israeli minority by the reigning Jewish hegemony). All systems fall under the umbrella of the Israeli Ministry of Education (Sprinzak et al., 2001). In spite of structural constraints reflected in the educational system, the Palestinian population in Israel has made great progress. For example, literacy rates, at present, reach over 90 percent of the population (Israel, 1995). Nevertheless, when compared to the Jewish school system, great discrepancies exist in terms of physical facilities, teacher qualifications, retention rates, and levels of special services (Rouhana, 1997).

The language of instruction in Jewish schools is Hebrew and in Palestinian schools Arabic. For Jewish schools, English and Arabic serve as second languages (with English as the preferred option). Within Palestinian schools Hebrew is now taught at the second grade and English is taught as a third language. At the University level, most Jews and Arabs attend school together, with the language of instruction being Hebrew (Al-Haj, 1995; Hertz-Lazarowitz, 1988). This constrains Palestinian rate of success.

Some of the features of the Palestinian educational system in Israel reflect the unique sociocultural background of this population (Abu-Nimer, 1999). Within the classroom, student-teacher relationships follow an authoritarian model and teachers employ a frontal teaching approach that is pedagogically very traditional. Teachers report a sense of conflict regarding their loyalty toward their employer, the Ministry of Education, and their loyalty toward their Palestinian community. The Ministry of Education imposes curricular constraints on the Palestinian educational system by, for example, not allowing schools to choose freely their own narratives concerning issues related to cultural and national histories. From a curricular perspective, little attention is paid to Palestinian history and culture though Palestinian students learn Zionist and Jewish history as well as Jewish literary studies (Rouhana, 1997). The security measures, traditionally used by Israeli (Jewish) officialdom to restrict teacher appointments, were canceled only in 1994 (Kretzmer, 1990; Rouhana, 1997). Consequently, an enormous gap has developed between the two systems, leaving the Palestinian-Israeli educational system decades behind.

**The Socio-Ideological Context of the Bilingual Initiative**

The present initiative is not the first attempt at bilingual, desegregated education in Israel (See Feuerverger, 2001; Gavison, 2000). The central difference between the first project, called the Neveh Shalom-Wahat al-Salam School, and the present initiative lies in the environments in which each has evolved. The Neveh Shalom-Wahat al-Salam School is situated in a small settlement, ideologically identified with the vision of full equality for Palestinian and Jewish community members. The bilingual programs initiated in 1998 by the Center for Bilingual Education in Israel (CBE) were implemented in mixed residential and urban areas without initial community support.

The idea of creating Palestinian-Jewish coeducation is, in and of itself, a daring enterprise. The CBE, established in 1997, sought to foster egalitarian Arab-Jewish cooperation in education, primarily through the development of bilingual and multicultural coeducational institutions (Bekerman and Horenczyk, 2004; Bekerman and Nir, 2006). In 1998, the center established two schools guided by these principles, one in Jerusalem and one in the Upper Galilee. A third school opened in 2004 in Kfar Karah, the first to be established in a Palestinian village thus revolutionizing basic Israeli perspectives that could accept integrated schools in Jewish majority settlements but had difficulty with the idea of sending Jewish children to a segregated Palestinian area.

At present the schools run 19 classes, from kindergarten to seventh grade. In the Jerusalem school, 122 Palestinian children (54 boys and 68 girls) and 107 Jewish children (59 boys and 48 girls) are enrolled. In addition 23 children fall into the category of "the other" (13 boys and 10 girls).

In Upper Galilee, there are eight classes in the school, from kindergarten to sixth grade, including two second grades. The school in the Upper Galilee (Misgav) is attended by 93 Palestinian children (56 boys and 37 girls) and 77 Jewish children (42 boys and 35 girls). There are 7 classes, from first grade to seventh grade, with 9 Palestinian children and 10 Jewish children in seventh grade. The school in Kfar Karah opened with 103 students: 53 Palestinian students (34 boys and 19 girls) and 50 Jewish students (28 boys and 22 girls) in 4 classes running from kindergarten to third grade.

The schools are recognized as nonreligious and are supported by the Israeli Ministry of Education. They use the standard curriculum of the state nonreligious school system, the only difference being that both Hebrew and Arabic are languages of instruction. The CBE educational initiative has to confront what Spolsky and Shohamy (1999) have characterized as a Type 1 monolingual society in which a sole language (Hebrew) is recognized as associated with the national identity while other languages (i.e., Arabic), though officially recognized as a second language for education and public use (Koplewitz, 1992; Spolsky, 1994), have been marginalized. In an attempt to offset this marginalization, the schools employ a strong additive bilingual approach emphasizing symmetry between both languages in all aspects of instruction (Garcia, 1997). The CBE documents posit that bilingual study can be instrumental in deepening each group's understanding of the other. It perceives bilingual education as empowerment pedagogy that helps increase the self-esteem of minority students.

From its inception, the CBE has adhered to an ideology emphasizing sustained symmetry at all organizational, curricular, and practical levels. They have successfully sustained this goal by securing the services of a well-balanced educational staff, having a Palestinian and a Jewish teacher in each class and having a Palestinian and a Jewish school principal. In terms of aims and processes, the initiators of the bilingual project would likely agree with Skutnabb-Kangas and Garcia's three main benefits of an effective bilingual educational project: (1) a high level of multilingualism; (2) equal opportunity for academic achievement; and (3) a strong, positive multilingual and multicultural identity including positive attitudes toward the self and the other (Skutnabb-Kangas and Garcia, 1995).

As expressed in its formal publications, the CBE aims to develop a new educational scheme where children, parents, and the larger community together with governmental institutions (Ministry of Education local authorities) build a cooperative framework structured on the basis of equality and mutual respect. In 2003, the CBE officially changed its name to the Center for Arab Jewish Education in Israel, possibly reflecting the need to emphasize educational integration and multiculturalism more than its bilingual goals.

## Methodology

The results reported in the following section, aside from those reporting on teachers' perspectives, correspond to research conducted in two of the three schools over a period of five years beginning in 1999. The latter include representatives of all three schools under the CBE umbrella. Research efforts continue and will transpire for at least two more years. Research was conducted using a variety of qualitative methods including participant observation, in-depth interviewing, and documentation gathering. Much of these results have been published in a variety of academic publications (Bekerman, 2003a , 2003b, 2004, 2005, Bekerman and Shhadi 2003; Bekerman and Maoz; 2005; Bekerman and Horenczyk 2004; Bekerman and Nir, 2006). Throughout the years, we have conducted over 100 interviews with parents, most in individual sessions lasting approximately an hour and the remainder in small group meetings lasting approximately 90 minutes. Most staff members were interviewed many times during the length of the research. We also spoke with students, either in brief semi-structured individual interviews or in more informal circumstances and conducted systematic interviews with the second and third grade cohort at the Misgav school (in 2000). Finally, we attended and recorded a variety of teacher, parent, and steering committee meetings. Observations were and continue to be conducted in classrooms that today account for over 100 days of school activity including special school and community events: field trips, ceremonies, festivals, and memorial days. Throughout the research process, the research team included both Palestinian and Jewish researchers who, for the most part, are fluent in both Hebrew and Arabic. Accordingly, all interviews were conducted in Hebrew or Arabic depending on the preference of the interviewee. Interviews and field observations were, in their majority, audio or video taped and transcribed. They were analyzed according to conventional qualitative methods (Mason, 1996; Silverman, 1993). We have monitored our interpretative efforts through peer debriefing paying special attention to the ways in which we, as researchers, allowed or did not allow for the preliminary coding to be influenced by our prior expectations or theoretical inclinations. We used negative case analysis to gain confidence in the hypotheses proposed. We carefully analyzed the data, looking for patterns and thematic issues of relevance, which were then coded as to allow for further analysis. High levels of agreement between the coders were reached after thorough discussions (Glassner and Loughlin, 1987). Thus we arrived at identifiable themes from which we created our final coding system. Throughout the process, intercoder reliability checks showed strong agreement between coders and high reliability for the coding scheme. Moreover, and in line with naturalistic critical perspectives (Carspecken, 1996), the final coding scheme was further checked for validity and reliability.

### Summary Results

#### Language

Like many bilingual programs, the bilingual schools studied suffer from somewhat contradictory practices, perspectives, and expectations in relation to their goals. Despite serious efforts by the entire staff, the attempt to sustain full symmetry through the implementation of what we have come to call the "rule of the ruler" failed. The double meaning of the term "ruler"—as both an empowered individual/group and a measuring device—implied the need to guarantee symmetry or equality when designing and implementing all school activities. Our observations show that upon entering any classroom, it becomes immediately apparent that one has entered a bilingual world. All signs, letters, and numbers hanging on the walls, the books in the classrooms and the library, and the trilingual computer keyboards (Arabic, Hebrew, and English), reflect effort invested in language equality. The "rule of the ruler," though a central means for recognizing two national/cultural traditions, could not alone overcome the realities of the mostly monolingual Israeli society. Even when the language policy shifted toward an "affirmative action" approach in support of Arabic, the introduction of English and Israel policies rendered the bilingual efforts mostly ineffective with regard to the Jewish population at school (at least for now).

Teachers expressed ambivalence toward the bilingual policy. Though they support it, they were conscious that much of what went on in the educational background remained biased toward the Hebrew language. All teachers interactions we observed and recorded were conducted in Hebrew since, for the most part, Jewish teachers have no Arabic literacy and Hebrew is the prevailing language in all staff meetings and training sessions as well as in meetings of parents and of the steering committee. All teachers and students who related to language use in their interviews indicated that interactions in class and recess among children of different national groups were conducted, but for few exceptions, in Hebrew. This pattern also emerges from all our videotaped and written observations.

Different motivations regarding language literacy support the participating groups. Jewish parents sometimes joined the schools in order to materialize their ideological aspirations, albeit under certain constraining conditions. For example, they support bilingualism as long as it does not harm educational excellence and seemed satisfied with an educational initiative that allowed them to substantiate their liberal positions and to offer their children cultural understanding and sensitivity toward the "other." Palestinian parents seemed to be after the best education available given the present Israeli sociopolitical context. As apparent from the interviews, Israel's present sociopolitical conditions make it almost impossible for parents to dream about a soon-to-arrive top-down multicultural multilingual policy. It is also not totally clear whether they would adopt for it if imposed, especially considering that, as members of an upper-middle socioeconomical sector of society, they perceive education to be a means of mobility in a world going global. They thus preferred an English lingua franca and high Hebrew literacy for their children's future.

Our research has exposed multiple macro and micro contextual levels embedded in bilingual education. At the macro national level, we find a still segregationist policy that offers no "real" support for multilingual policies. At a medium community level, we see competing in-group perspectives and cross-group partial coalitions. At the level of educational practices, we see an outstanding effort to achieve dual bilingual opportunities for all, first by emphasizing symmetry between the languages and later by explicitly preferring Arabic over Hebrew in order to attain the program's objectives.

Despite tremendous effort and intention, these initiatives seem, in the best cases, not to be attaining their goals and, in the worst cases, oblivious to the reasons for their failure. At present, the bilingual project seems primarily to serve political agendas within the realm of the nation-state in a world where national boundaries no longer represent clear-cut national identities and are affected, at least in the West, by large waves of voluntary and involuntary immigration. At times, these agendas strive sincerely to promote the interests of minority groups and confront mainstream hegemonies. At other times, such agendas simply pay lip service to political correctness.

We could easily fault the teachers and parents involved in our program. We could blame them for consciously or unconsciously conveying negative messages about the minority language in spite of their overt efforts to create a school environment and a curriculum that represents a balanced bilingual effort. But this would be clutching at straws. If placed anywhere, "blame" should be placed on an adaptive, wider, sociopolitical system in which Arabic carries little symbolic power. In Bourdieu's (1991) terms, in Israel speakers of Hebrew have more cultural capital in the linguistic market place than those who speak Arabic.

It is not clear whether the parents participating in the initiative are interested in changing the existing power relations in Israel. The Jewish parents, reflecting the societal majority, while clearly liberally inclined and hopeful in creating more humane and respectful environments for the Palestinian-Israeli minority, do not necessarily see a need for radical change. The Palestinian parents, who belong to an aspiring middle class, understand the advantages of linguistically empowering their children and adapt to the rules of the game, specifically in a school context that, at least declaratively, stands behind an emancipating option.

We cannot assume that solutions to these issues can be found within the narrow limits of the schools and their surrounding communities. What undoubtedly needs to be addressed are the deeply entrenched paradigmatic perspectives that support the nation-state ideology and its traditional monoculturalism and monolingualism.

## Religious, Cultural, and National Identity

When compared to national issues, cultural and religious issues seem relatively easy. Our research shows that parents and teachers see culture and religion as areas in which mutual understanding can help bridge the gaps that separate both populations in Israel. Parents stress getting to know and understand the others' culture better and believe that the schools are achieving this goal . Teachers emphasize similar issues and educational activities around these issues appear to be conducted with ease and in fruitful collaboration. Cultural and religious issues become more salient during special events such as school trips (e.g., visits to a nearby synagogue, mosque, and church), or festive events such as the celebration of Hanukkah, Id Fitter, and Christmas that allow for broad expressions of solidarity and mutual understanding.

These celebrations carry a strong religious emphasis. In fact, it could be said that religious aspects are disproportionately emphasized given that the majority of the Jewish parents belong to the secular segments of Israeli society and the Muslim populations, though more traditionalist are mostly nonreligious. All three schools allocate special time to religious studies, which are generally conducted separately for the Jewish, Muslim, and Christian populations. At times, Jewish parents express concerns and ambivalence about this religious emphasis. At the same time, they seem to find solace in the religious underpinning given their mostly unarticulated fear that their children's Jewish identity will erode as a result of participation in a binational program. Jewish teachers seem to share this attitude emphasizing knowledge of Jewish traditions as an antidote to the perceived superficiality of secular Jewish identity and its possible weakening within a binational environment.

At the cultural religious level, school communities are invited to share a common sphere. At times these commonalities supersede the minority tradition. For example, in the first years of the schools' activities, the traditional Jewish festival of the "reception of the Torah" served as an opportunity to include the events of receiving the Koran in the Moslem tradition thus circumscribing the traditions by their commonality but paradoxically negating the very principle of respect and recognition upon which the integrated schools stand. Nevertheless, the staff involved in this refreshing educational adventure continuously experiments with new approaches and has in recent years dealt with some of the difficulties experienced in the past.

The ethnographic data suggest that issues of national identity have become the ultimate educational challenge for parents and educational staff alike. National issues are compartmentalized into a rather discrete period in the school year corresponding in the Jewish Israeli calendar to Memorial Day and subsequently Israel's Independence Day and in the Arab calendar to the Day of the Naqbe. In accordance with the policy of the Ministry of Education, all schools hold a special ceremony for the Jewish cohort on Memorial Day that the Palestinian cohort does not attend. Depending on the schools (complex) relations with the surrounding community and the Ministry of Education's supervision, a separate ceremony is conducted for the Palestinians in remembrance of the Nakbe. If not possible or decided otherwise, the Nakbe is commemorated on a different day and or in an extracurricular activity. During the first year of our study, under the optimism following the Oslo agreements, these events progressed rather smoothly. During the second year, 2000, these events were strongly influenced by the tense and violent political climate following the Land Day calamity in which 13 Palestinians were killed by Israeli police forces. Much of the educational work during the second year ensued under a growing sense of suspicion from both national groups represented at the schools.

On the Jewish side, these tensions have somewhat relaxed. The schools are more established and Jewish parents, well guarded by the Ministry of Education rulings regarding Memorial Day ceremonies, are confident that from their perspective the schools support their national needs. For the Palestinian group, tensions continue, particularly among the teachers, who see themselves at the forefront of the struggle to safeguard the Palestinian national narrative that remains unrecognized by the Israeli educational officialdom. Though Jews at the schools clearly represent the politically liberal, center–left segments of Israeli society, at times Palestinian national expression does not always fall within the limits of legitimate expression as delineated by the liberal Jews. For most liberal Jews Israeli Palestinian cultural and religious expression in school is legitimate. However, national identification with the Palestinian authority is not welcomed. An exception to this emerged when Yasser Arafat, the president of the Palestinian authority, died. After an emergency, parents' meeting was held, one of the schools decided to hold an event commemorating Arafat's death that included mounting a remembrance tableau similar to and in the place where Yithak Rabin, the assassinated Israeli prime minister, was remembered a few days earlier.

The developing bilingual schools' experience teaches us valuable lessons about the flexibility of religious and national symbols and about the complex give-and-take processes involved in intergroup cooperative work. The bilingual schools represent a commendable effort toward a more respectful and humane relationship between ethnic groups in a conflict-ridden area. Working partially against the state-imposed and ideologically normative Jewish hegemony, the participants in the initiative search for ways to counter the weight of ideology and tradition. In spite of these ongoing efforts, the educational initiative's approach to national and cultural identity, perhaps most succinctly expressed in ceremonies and festivals, seems to move along an essentialist continuum from a somewhat accommodating religious/cultural identification to a stricter national one, all under the banal ideology of the nation-state. In a sense, the religious cultural side of the continuum allows for more flexibility, experimentation, and creativity. On the opposite side, we see that national ceremonies risk endangering the initiative in its totality.

Our research demonstrates the need to shift the focus of multicultural education away from ceremonial activity and toward long-term educational efforts dispersed throughout the school year and across multiple disciplines. Our research also indicates the need to move away from categories left vacant in the discursive resources offered by the context of the nation-state and toward the unveiling of other available resources and their shaping forces. The eurocentric maladies multiculturalism set out to abolish are intrinsic in nation-state ideology. As such, only a critical approach toward the epistemological bases of the nation-state can help to overcome them. This study advocates the need for multicultural efforts free from the hegemonizing power of the universality of national ideology. There are no universal multicultural approaches independent of sociopolitical contexts. Every multicultural endeavor involves new imaginings and difficult, hazardous work.

## Social Interaction and Children's Perspectives

"Nobody is really happy" is a statement that reflects what adults involved in the bilingual schools feel regarding social interactions. This does not mean they are upset;

they just expected more. During the first two years of our research, classroom intergroup interaction was greater than during recess or other unstructured periods. A recurring pattern observed during the recess periods showed Jewish and Arab children playing separately: the Jewish children played tag while the Arab children mostly played soccer. In conducted interviews, most children were aware of this pattern but did not seem to regard it as a problem. Sometimes they explained it in terms of personal preferences (e.g., Jewish children do not like to play soccer). In class, children worked together in cross-national teams and assisted each other with different assignments, but when at home, intergroup visits were rare. Throughout the years of our research, we have seen some progress and today in the higher grades, interactions are more frequently recorded.

All in all the bilingual schools seem to be partially successful in helping to reduce prejudice and alleviate conflict. These findings arise from our continued effort to interview both students at the Misgav bilingual school (starting with second and third graders in 2000) and, for comparative purposes, interviews with children studying in parallel grades at the Sachnin Palestinian School and the Misgav Jewish School, both state monolingual schools in Israel. The responses of children in the bilingual schools to questions related to political/conflictual events are in general more moderate than those expressed by children in the regular monolingual schools. Moreover, from an analysis of their responses to questions related to cultural/religious matters, it is apparent that the children's understanding of one another's cultures runs deeper in the bilingual settings. Central to this success is the opportunity for close, sustained, and cooperative contact in a context that is anxiety reduced and equitable for all participating groups. Like students attending the monolingual schools, bilingual school participants still recognize themselves as ethnically/religiously/nationally divergent. However, they differ from students in monolingual schools in that they express less of a sense of social distance between the groups. Whether these positive effects are transferable outside the immediate educational environment remains unknown. We need sustained research efforts to uncover the sustainability of these effects in the future when/if these children join mainstream educational tracks and move into adulthood.

## Conclusions and Challenges

The bilingual schools is a product of the entrepreneurship of good willing citizens who, unaided by theoretical conceptualizations, developed a system based on commonsensical humane approaches flowing from their experience in the complex and unjust Israeli society. Theoreticians and researchers should be humbled by this simple fact and struggle in their work to be attentive, make sense, and try to better understand this folk knowledge.

The schools' functioning can be conceptualized through a variety of existing theoretical paradigms including peace and coexistence education (Bar-Tal, 2002), collaborative learning (Slavin and Cooper, 1999), multiculturalism (Banks, 1995), bilingual education (Delpit, 1988; Hornberger, 2002), contact hypothesis (Allport, 1954), and though less recognized acculturation theories (Berry, 1997). These paradigms can be relevant to the development of better strategies and pedagogies that can help to overcome mistrust and fear between the communities involved in this educational initiative. Central to this endeavor is the effort researchers will need to invest, in order to create the necessary conditions in educational surroundings to have their voices heard; not at the relatively easy policy level, but with teachers who stand at the only real front of this effort. They are not usually supported enough to be attentive to analytical systematic thinking while conducting their very complex activities.

Dealing with the totality of the potential cross-fertilization between different theoretical perspectives that can support the schools' work will be left for future writings. Space limitations prevent us from elaborating at this turn. Nevertheless, I want to conclude by pointing out two directions that I believe are central to the future development of the educational initiatives under study.

Research has shown that intergroup contact seems to generally promote intergroup acceptance, especially when appropriate conditions for contact are met (Pettigrew, 1998; Pettigrew 1998 and Tropp, 2000).Our ongoing research shows the potential benefits of one type of intergroup contact, namely, bilingual long-term coeducation, and also sheds light on the complexity and difficulties facing all the parties involved in such an adventurous enterprise. It is important to note that intergroup contact can emphasize various—and sometimes contradicting—strategies: decategorization (or personalization), categorization, crossed-categorization, and recategorization (Hewstone, 1996). In the bilingual schools described in our study, neither decategorization nor crossed-categorization approaches are utilized, thus preventing the implementation, at least partial, of strategies that allow for an increase of complexity in intergroup perceptions. The schools seem to adopt a purely categorized approach, based on the premise that strengthening ethnic and national identities is the path to achieving its aims. The adoption of a categorized approach needs to be critically considered and these considerations are, at this point, lacking in the schools' discourse.

Though perceived as the solution to parents' fears, categorization is not necessarily the ideal direction if what we are after is serious mutual (group and individual) recognition and respect. Categorization builds on the imposition of preaccepted categories and often limits freedom and adaptation not to speak of change. Categorization supports political interests and in Israel these generally support the Jewish majority and the inequalities their system has developed. Recategorization, which calls for the development and strengthening of a "common ingroup identity" (Gaertner, Dovidio, Anastasio, Bachman, and Rust, 1993), does not necessarily mean enhancing the perception of common "Israeliness," which, given the present contextual constrains, might be understood as a means for replicating the present sovereign Jewish scheme. However, it could be interpreted as fostering an overarching civil identity whose contents have yet to be renegotiated.

When taking this path we need be aware that contact theories mainly deal with prejudice reduction rather than interactional societal developments among individuals from groups in conflict. The schools deal with living individuals who, in their own way, consider group relations (which as our study shows at times shadow the initiatives potential) and individual relations. Similar issues have been raised in coexistence theorizing. Kriesberg (1998), for example, is well aware that coexistence is an open concept that "leaves a great deal of room for various forms of relations" (p. 183). Bar-Tal (2004) recognizes the vagueness and indistinctiveness of the concept and the fact that it pertains only to minimal positive intergroup relations.

The very fact that schools are involved in daily activities characterized by high social proximity suggests that they are in need of much more. Schools deal only with individual intergroup relation saturation. They need to make clear (or at least delineate the parameters) of their policies regarding this intergroup individual reality. Articulating these policies can help alleviate many of the tensions felt by all participants in the involved community. In short, we should question if "separate can be equal." As indicated by Hewstone (1996), distinct approaches such as categorization, decategorization, or recategorization are likely to lead to different outcomes. We believe that they may also present program developers and practitioners with different challenges and problems.

For a while I have considered complementing contact theories with acculturation theories in line with the fourfold model delineated by Barry (1997) that, regardless of the culture, topic, or intent, posits four generic types of acculturation: assimilation, separation, integration, and marginalization. Barry's fourfold model seems to fall into a trap similar to those related to contact perspectives. Dependence on reified conceptions of culture and identity does not reflect the complexity and potential outcomes of individual interaction. Rudmin has successfully posited that integration construct, while plausible for surface behaviors amenable to code-switching, such as cuisine, language, and music, is not amenable for more defining cultural aspects not open to code-switching such as religion, sexual norms, cleanliness, child-rearing, and so on (Rudmin and Ahmadzadeh, 2001, p. 43). In these contexts, biculturalism is limited if not impossible. Though acculturation theory seems not to allow for much progress regarding the schools' needs in terms of intergroup educational policy articulations, it points in a welcomed direction in terms of the specifics of intergroup contact. As such, it might be useful to consider when trying to further contact theorizing.

The bilingual schools challenge our theoretical imagination. They compel us to consider specific individual experiences as well as individuals and their group affiliation. Cross-fertilization between present paradigmatic perspectives and their better adaptation to individual behaviors in all their complex details might be the direction we need to take if we want to contribute to what the bilingual schools are already doing.

The bilingual schools are indeed an extraordinary new experiment, one that carries the promise of a better future for Jewish–Palestinian relations in Israel. Their work is done on roads not yet traveled and—although always at risk—they carry the dream for a peaceful and more honorable world. It is yet to be seen if the foundational ideology that sustains them can survive the present escalating conflict and achieve its goals.

### Note

This chapter is based on a project funded by grants from the Ford Foundation, Grant Number: 990–1558, the Spencer Foundation, and the Bernard Van Leer Foundation. We are very grateful to the staff at the Center for Bilingual Education in Israel and to the principals and teachers of the three bilingual schools for their cooperation and encouragement.

### References

Abu-Nimer, M. (1999). *Dialogue, conflict, resolution, and change: Arab-Jewish encounters in Israel.* Albany: SUNY.

Al-Haj, M. (1995). *Education, empowerment, and control: The case of the Arabs in Israel.* Albany, NY: SUNY Press.
Allport, G.W. (1954). *The nature of prejudice.* London: Addison-Wesley.
Anderson, B. (1991). *Imagined Communities: Reflections on the Origins and Spread of Nationalism.* London: Verso.
Banks, J.A. (1995). Multicultural education; Historical development, dimensions, and practice. In J.A. Banks and C.A.M. Banks (Eds.), *Handbook of research on multicultural education* (pp. 3–24). New York: Macmillan.
Barakat, H. (1993). *The Arab world: Society, culture and state.* Berkeley, CA: University of California Press.
Bar-Tal, D. (1990). *Group beliefs: A conception for analyzing group structure.* New York: Springer-Verlag.
———. (1998). Societal beliefs in times of intractable conflict: The Israeli case. *International Journal of Conflict Management,* 9: 22–50.
———. (2002). The elusive nature of peace education. In G. Salomon and B. Nevo (Eds.), *Peace education* (pp. 27–36). New Jersey: Lawrence Erlbaum.
———. (2004). Nature, rationale, and effectiveness of education for coexistence. *Journal of Social Issues,* 60(2): 253–271.
Bekerman, Z. (2003a). Never free of suspicion. *Cultural Studies <> Critical Methodologies,* 3(2): 136–147.
———. (2003b). Reshaping conflict through school ceremonial events in Israeli Palestinian-Jewish co-education. *Anthropology & Education Quaterly,* 34(2): 205–224.
———. (2004). Multicultural approaches and options in conflict ridden areas: Bilingual Palestinian-Jewish education in Israel. *Teachers College Record,* 106(3): 574–610.
———. (2005). Complex contexts and ideologies: Bilingual education in conflict-ridden areas. *Journal of Language Identity and Education,* 4(1): 1–20.
Bekerman, Z. and Horenczyk, G. (2004). Arab-Jewish bilingual coeducation in Israel: A long-term approach to intergroup conflict resolution. *Journal of Social Issues,* 60(2): 389–404.
Bekerman, Z. and Maoz, I. (2005). Troubles with identity: Obstacles to coexistence education in conflict ridden societies. *Identity,* 5(4): 341–358.
Bekerman, Z. and Nir, A. (2006). Opportunities and challenges of integrated education in conflict ridden societies: The case of Palestinian-Jewish schools in Israel. *Childhood Education,* 82(6): 324–333.
Bekerman, Z. and Shhadi, N. (2003). Palestinian Jewish bilingual education in Israel: Its influence on school students. *Journal of Multilingual and Multicultural Development,* 24(6): 473–484.
Ben-Amos, A. and Bet-El, I. (1999). Holocaust day and Memorial day in Israeli schools: Ceremonies, education and history. *Israel Studies,* 4 (1): 258–284.
Ben-Rafael, E., Olshtain, E., & Geijst, I. (1994). *Aspects of Identity and Language in the absorption of the immigrants from the FSU.* Jerusalem (Hebrew): Institute for Educational Research, Hebrew University.
Berry, J.W. (1997). Immigration, acculturation, and adaptation. *Applied Psychology: An International Review,* 46: 5–68.
Bourdieu, P. (1991). *Language and symbolic power.* Cambridge: Harvard University Press.
Carspecken, P.F. (1996). *Critical ethnography in educational research: A theoretical and practical guide (critical social thought).* New York: Routledge.
Delpit, L.D. (1988). The silenced dialogue: Power and pedagogy in educating other people's children. *Harvard Educational Review,* 58 (3): 280–298.
Feghali, E. (1997). Arab cultural communication patterns. *International Journal of Intercultural Relations,* 21: 345–378.
Feuerverger, G. (2001). *Oasis of dreams: Teaching and learning peace in a Jewish-Palestinian village in Israel.* New York: RoutledgeFalmer.

Gaertner, S.L., Dovidio, J.F., Anastasio, P.A., Bachman, B.A., and Rust, M.C. (1993). The common ingroup identity model: Recategorization and the reduction of intergroup bias. In W. Stroebe and M. Hewstone (Eds.), *European Review of Social Psychology* (vol. 4, pp. 1–26). Chichester: Wiley.
Garcia, O. (1997). Bilingual education. In F. Coulmas (Ed.), *The handbook of sociolinguistics* (pp. 405–420). Oxford: Blackwell.
Gavison, R. (2000). Does equality require integration? *Democratic Culture*, 3: 37–87.
Gellner, E. (1997). *Nationalism*. New York: New York University Press.
Ghanem, A.A. (1998). State and minority in Israel: The case of ethnic state and the predicament of its minority. *Ethnic and Racial Studies*, 21 (3): 428–448.
Glassner, B. and Loughlin, J. (1987). *Drugs in adolescent worlds: Burnouts to strights*. New York: St. Martin's Press.
Handelman, D. (1990). *Models and mirrors: Towards an anthropology of public events*. Cambridge: Cambridge University Press.
Hertz-Lazarowitz, R. (1988). Jews and Arabs in conflict on campus: A social drama perspective. In J.E. Hofman (Ed.), *Arab-Jewish relations in Israel* (pp. 271–301). Bristol, UK: Wyndham Hall.
Hewstone, M. (1996). Contact and categorization: Social psychological interventions to change intergroup relations. In C.N. Macrae, C. Stangor, and M. Hewstone (Eds.), *Foundations of stereotypes and stereotyping* (pp. 323–368). New York: Guilford.
Hobsbawm, E.J. (1983). The invention of tradition. In E.J. Hobsbawm and T. Ranger (Eds.), *The invention of tradition* (pp. 1–14). Cambridge: Cambridge University Press.
Hofstede, G. (1980). *Culture's consequences*. Beverly Hills, CA: Sage.
Hornberger, N.H. (2002). Multilingual language policies and the continua of biliteracy: An ecological approach. *Language Policy*, 1: 27–51.
Kelman, H.C. (1999). The interdependence of Israeli and Palestinian national identities: The role of the other in existential conflicts. *Journal of Social Issues* (55), 581–600.
Koplewitz, I. (1992). Arabic in Israel: The sociolinguistic situation of Israel's linguistic minority. *International Journal of the Sociology of Language*, 98: 29–66.
Kretzmer, D. (1990). *The legal status of the Arabs in Israel*. Boulder, CO: Westview Press.
Kriesberg, L. (1998). Coexistence and reconciliation of communal conflicts. In E. Wiener (Ed.), *The handbook of interethnic coexistence* (pp. 182–198). New York: Continuum Publishing Company.
Mason, J. (1996). *Qualitative researching*. London: Sage publications.
Pettigrew, T.F. (1998). Intergroup contact theory. *Annual Review of Psychology*, 49: 65–85.
Pettigrew, T.F. and Tropp, L.R. (2000). Does intergroup contact reduce prejudice? Recent meta-analytic findings. In S. Oskamp (Ed.), *Reducing prejudice and descrimination* (pp. 93–114). Mahwah, NJ: Earlbaum.
Rouhana, N. (1997). *Palestinian citizens in an ethnic Jewish state*. New Haven: Yale University Press.
Rudmin, F.W. and Ahmadzadeh, V. (2001). Psychometric critique of acculturation psychology: The case of Iranian migrants in Norway. *Scandinavian Journal of Psychology* (42): 41–56.
Said, E. (1994). *The Politics of dispossession*. New York: Vintage.
Silverman, D. (1993). *Interpreting qualitative data*. London: Sage.
Skutnabb-Kangas, T. and Garcia, O. (Eds.). (1995). *Multilingualism for all? General principles*. Lisse: Swets and Zeitlinger.
Slavin, R.E. and Cooper, R. (1999). Improving intergroup relations: Lessons learned from cooperative learning programs. *Journal of Social Issues*, 55 (4): 647–663.
Smooha, S. (1996). Ethno-democracy: Israel as an archetype. In P. Ginosar and A. Bareli (Eds.), *Zionism: A contemporary polemic* (pp. 277–311). Jerusalem: BenGurion University (Hebrew).
Spolsky, B. (1994). The situation of Arabic in Israel. In Y. Suleiman (Ed.), *Arabic sociolinguistics: Issues and perspectives* (pp. 227–236). Richmond: Curzon Press.

Spolsky, B. and Shohamy, E. (1999). Language in Israel society and education. *International Journal of the Sociology of Language*, 137: 93–114.

Sprinzak, D., Segev, Y., Bar, E., and Levi-Mazloum, D. (2001). *Facts and figures about education in Israel.* Jerusalem: State of Israel, Ministry of Education, Economics and Budgeting Administration.

Abdo, N. and Yuval-Davis, N. (1995). Palestine, Israel, and the Zionist settler project. In D. Stasiulis and N. Yuval-Davis (Eds.), *Unsettling settler societies: Articulations of gender, race, ethnicity, and class* (pp. 291–322). Thousand Oaks, CA: Sage.

Khalidi, R. (1994). Toward a solution. In *Palestinian refuges: Their problems and future.* Washington, DC: The Center for Policy Analysis on Palestine.

Kimmerling, B. and Migdal, J. (1993). *Palestinians: The making of a people.* Cambridge, MA: Harvard University Press.

CBS (1995). *Statistical abstract of Israel.* Jerusalem: Central Bureau of Statistics.

Ghanem, A. (2001). *The Palestinian-Arab minority in Israel 1948–2001: A political study.* Albany, NY: SUNY Press.

Rekhess, E. (1988). First steps in the crystallization of Israeli policy toward the Palestinian minority. *Monthly Review.* 33–36.

Rouhana, N. and Ghanem, A. (1998). The crisis of minorities in ethnic states: The case of the Palestinian citizens in Israel. *International Journal of Middle East Studies* (30): 321–346.

Suleiman, R. (2004). Planned Encounters between Jewish and Palestinian Israelis: A Social-psychological perspective. *Journal of Social Issues*, 60(2): 323–337.

Smooha, S. (2004). *Index of Arab-Jewish relations in Israel.* Haifa: University of Haifa, The citizens' accord forum, The Jewish-Arab Center, Friedrich Ebert Stiftung.

Stasiulis, D. and Yuval-Davis, N. (1995). *Unsettling settler societies: Articulations of gender, race, ethnicity, and class.* Thousand Oaks, CA: Sage.

# Chapter Eight
## Peace Education in a Bilingual and Biethnic School for Palestinians and Jews in Israel: Lessons and Challenges

*Ilham Nasser and Mohammed Abu-Nimer*

### Introduction

#### Bilingual Schooling and Education for Peace

Peace education involves a slow process of change as opposed to an end product; therefore its impact is not tangible and easily quantified. As an emerging field of study, authors and editors of a recent special edition of the *Journal of Peacebuilding and Development* on peace education agreed that it is much easier to detect or trace the impact of peace education programs on the micro (individual especially the direct, immediate, and short-term impacts) than the macro level.[1] Changes in students' and teachers' attitudes are detected through surveys and other measurement forms. However, the larger impact of these programs is hardly traceable and few scholars have attempted such longitudinal research designs.

Similarly, the impact of bilingual education on attitudes toward peace and conflict has not been fully examined and the bilingual and integrated education experiences in conflict areas are even less addressed. There are several studies on bilingual education and its impact on relations between minority and majority groups but there is a scarcity of research on bilingual education as a tool for peace education and conflict management (Bekerman, 2004). It is, however, documented that bilingualism can bring about greater understanding among groups and an increased knowledge of each other (Bekerman, 2003; Garcia, 1997; Feuerverger, 2001). In fact, Garcia claims that "bilingualism creates a greater understanding that is beyond multicultural education and anti-racist education movement" (1997, p. 406).

Bilingual education can also serve as an effective empowerment tool for indigenous groups. Different models of bilingual education achieve this goal to a certain degree while others maintain the status quo. The best results are attained when the minority language has a greater weight to balance the dominance of the majority language. On the other hand, models that offer minority languages as lessons in the curriculum sustain the unequal status quo between the two languages (Garcia, 1997).

Teachers are often resistant to the application and integration of cultural diversity in their curriculum (Bekerman, 2003; Feuerverger, 2001; Garcia, 1997). Such a challenge is increased in contexts where teachers become fearful of any political discussion or emotional outburst when conflict issues are raised by students. Bringing students from both sides increases the potential for confrontation between them and avoidance by teachers who lack the skills to deal with the emotions and attitudes expressed by their students.

Glazier's (2004) in-depth description of the coteaching experience in a bilingual school in Israel illustrates the type of intensive preparation such schools in conflict areas have to adopt in order to effectively address the contextual factors. The primary challenge in such enterprise is overcoming the self bias of the teachers themselves. Such a process requires teachers to examine his/her attitudes toward the other and the conflict in general as well as his/her role as an educator for peace. Such preparation is not only essential for language, history, religion, and other social studies teachers, but for all staff.

The long-term impact of these attempts on students and their attitudes is yet to be fully explored, nevertheless it is documented that when schools intentionally bring students from both sides to learn in a facilitated environment a positive change in perceptions takes place (Bekerman, 2004; Glazier, 2004). Glazier (2004) argues that the goal of these schools should not only be to reduce prejudice but also to create cultural fluency between both people (for Glazier, cultural fluency means "the ability to step back and forth between two cultures in which prejudice reduction is one part" (p. 142).

The contact hypothesis theory (Allport, 1954) that is the basis for many of the initiatives in prejudice reduction work emphasizes the "nature of the contact" and its effect as dependant on the kinds of people and environments involved. More recent examination of this theory (Cook, 1978; Pettigrew, 1986) asserts that the participants should be of equal status within the contact and interdependent to achieve common goals. Critics of this view (Hewstone and Brown, 1986) see that contact hypothesis is narrow and limited and that proper conditions should be created for successful intergroup contact such as bringing radical social change in society. This chapter investigates the views of Palestinian and Jewish Israeli teachers and students about the uniqueness and characteristics of their bilingual biethnic school.

## Arab Education System in Israel

The Palestinian Israeli minority constitutes about 18 percent of the citizenship holding population. Their status has been characterized by many researchers as a group who is caught between its national identification with the Palestinian people and their plight for independence and between its civic belonging to the State of Israel. Many members of this community, especially second generation citizens born in Israel have internalized certain aspects of civic life in addition to their solidarity with the Palestinian people. Despite the intense Israelization process experienced (such as mastering Hebrew and familiarity with Israeli Jewish culture) the majority of this population continues to experience economic, political, and educational policies of discrimination and deprivation.[2]

In the educational context, Israel has two separate education systems for Palestinian and Jewish Israelis. Historically, the governments of Israel created an "Arab office of education" that handled the curriculum, budgets, and hiring of staff and teachers that is separate from the rest of the school systems in the country.

Since the creation of the state, integrating the Arab school system in Israel has been a complex endeavor not only due to the geographical and residential separation, but mainly because of the Jewish definition of the state and "state security" considerations. Policies to ensure the Jewish identity of the state has dictated many policies developed to manage the Arab education system, whereby curriculum and hiring are subject to approval by Israeli Jewish supervisors. The security establishment continues to intervene in many aspects of the Arab school system. Also, because there were few educated Arabs who remained in Israel after 1948, the school system was, for many decades, run by a traditional leadership that struggled to provide basic educational services and was mainly occupied with survival concerns until early 1970s, when a new generation began entering the school system and new revisions were implemented by the Ministry of Education. This historical reality has resulted in a continuing desire among many Palestinian Israelis to seek alternative school systems in which their children can be educated in a more open, nationally and culturally nurturing environment.

Today and after decades of conflict, there are four bilingual biethnic schools in Israel. These are the only pioneer attempts in this country to educate for peace by bringing Palestinian and Jewish Israelis to learn and teach together. The NS/WAS is the first of these four schools and was created in 1983.[3] This primary school (preschool through eighth grade) has been carrying the torch for pioneer work in peace education in a general environment in the country that is hostile toward such efforts. The school is hosted by a pioneer Palestinian and Jewish community that, in-mid 1970s, launched the first joint village in Israel.

The question of how successful these schools have been in promoting peace and understanding among children and teachers has not yet been fully examined. The purpose of this chapter is to shed some light on the experiences of the first school to implement such a bilingual program in Israel and to examine its success and areas for improvements in implementing a peace education program that can become a model for the rest of the country and region. Here, the focus is on children and teachers' perspectives about peace, bilingual education, and uniqueness of the school. The chapter also addresses the staff's effort to educate for peace and create a unique curriculum for Palestinian and Jewish children.[4]

## Contextual Factors

The research began in 2004 and was completed during the spring of 2005. The Palestinian second Intifada was still capturing the political news and affecting relations between both people in Israel. In that context, there were many suicide bombings in Israeli cities and towns and the Israeli military continued its operations against Palestinians militia and civilians. The mobility of Palestinians in the occupied territories was severely restricted and tension continued to characterize Palestinian-Jewish relations in Israel.

Recent years also marked an increase in economic hardship in Israel where there were several budget cuts that mainly affected the poor, senior citizens, and schools. The NS/WAS lost the ministry's contributions in funding children's transportation that is the highest cost for the school. Also, there was a global decrease in funding such initiatives by donors, especially in the United States of America due to shifts in political priorities after September 11.

A shortage in qualified bilingual education teachers and programs in Israel has been a continuous problem especially since the Jewish and Palestinian systems operate almost in total isolation (even mixed cities like Lydda-Ramlleh run two separate education systems). There is also a lack of interest by the current government to support such programs even though Arabic is considered to be an official language in the country. This reality impacts on the pool of teachers to choose from and contributes to a general disregard of the importance of knowing Arabic in a country where a fifth of the population speaks Arabic as their first language.

These and other factors are strong forces against any bilingual school attempts in the country. Thus a great deal of the NS/WAS school's effort was directed toward surviving the everyday challenges that such a school faces. In fact, as indicated by many of its members, its mere existence under these harsh political and contextual realities is an important achievement by itself.

In 2001 the village decided to include children from the surrounding communities to increase the number of students and to provide an alternative education model to the neighboring towns and villages. Many children were recruited that resulted in a rapid and diverse expansion. Apparently this overwhelmed the school's student body and community since the teachers and administrators were not ready to handle the tension, intensity, and pedagogy of teaching children who reside outside the village (some of whom have special needs). This situation created new challenges to the growing school such as: more Ministry of Education pressure, turn over in teaching staff and administration, professional development and qualifications challenges, and the school's struggle to locate its model in bilingual education.

### Qualities of a Bilingual and Biethnic School/ Conditions for Success

All teachers and staff except for two were interviewed at the school (20 teachers, 8 administrators) in addition to 29 children. They all had a diverse view of the school and its unique approach in bilingual and peace education. Teachers and staff indicated a high level of engagement and emotional involvement in the school and its mission. Children enjoyed the freedom and opportunities to learn two languages and to meet children from the other side.

When staff was asked about the strength of the school and the qualities required of teachers who work there, several patterns emerged. In general there was an agreement that teachers should have qualities important for dealing with the "other," such as patience, respect, acceptance, tolerance, and ability to listen. More specifically, both Palestinians and Jewish Israeli teachers agreed that to "teach in such a school one needs to believe in the unique idea of the school." For example, Palestinian and Jewish teachers agreed that the "existence of such a school breaks a huge barrier between the two people." One said, "You have to believe in dialogue to work here! You should be able to accept others without limitations!" Others pointed out that "You have to believe that Arabs and Jews are equal and that they can learn and teach together."

Some, especially Palestinians, see themselves as agents to bringing change between Arabs and Jews in the larger society while others are skeptical of that. One Jewish teacher concluded: "I like the fact that both people learn and teach together, however I doubt how much impact we have on the larger communities." Still, instilling peace values was important because as one teacher stated, "the school exposes kids to a wider perspective and to building strong relationships." Several explained that children at the school are "learning not to fear the other side, learning to respect and accept, and learning to change stereotypes about the other." The joint space created in school nurtures values of coexistence, tolerance of the other, and allows interested teachers and children to pursue peace education processes.

Similar to Palestinians, the Israeli Jewish teachers enjoyed the school's cultural traditions and celebrations, which they cannot experience anywhere in their Jewish environment. Secular and religious Jewish schools only celebrate Jewish religious and cultural holidays and have no opportunity to experience Arab culture or tradition. While Palestinian Israeli study Jewish and Zionist movement history they have no opportunity to experience or know how Israeli Jews celebrate their cultural and religious holidays. The school also provides an opportunity to learn about different cultures from within the same ethnic group. One teacher explained, "Palestinian children come to us from many different places and different religions; we all get the chance to learn about each other and the Jewish culture."

Qualifications and educational approach were other benefits and requirement to succeed at the school. All agreed that teachers should have an alternative approach to teaching, communication skills, and mostly the ability to listen and understand children. Most teachers mentioned the need to have a strong educational background and teaching ability, to be sensitive and know how to manage a classroom, and to have the space to grow professionally. Several Palestinian teachers added: "Teachers who work at such a school should be flexible and able to accept change." All Palestinians insisted that teachers should be bilingual to teach in such a school while Jewish teachers were divided on that. One should mention that the school's written policies require teachers to be bilingual but due to the rapid expansion these policies were not implemented in recent recruitment of Jewish teachers.

The small size of the school also provided a flexible schedule and freedom for teachers to decide on what and how to teach. Several Palestinian teachers appreciated and particularly identified the opportunity offered by the school structure to know Jewish teachers in depth and learn from them. One teacher said: "We learn from the Jewish staff, they are more open in teaching." Overall, Jewish teachers' responses were focused more on the school's environment. Many described the school as warm, friendly, and small. However, most added that it was a unique and professionally challenging environment.

Interestingly, in addition to the benefits mentioned, all Palestinian teachers interviewed focused on strengthening self-confidence as a special benefit the school offers to Palestinian children. The school also increases Palestinian students' awareness of their identity, and teaches them critical thinking, creativity, and self-exploration. This emphasis on self confidence may be due to teachers' perception of their students as part of the minority group that obviously affects their ability to gain the same degree of confidence and self-esteem as their Israeli Jewish peers. Living as a minority in a Jewish state and the reality of intense security surveillance and tight control system, adults and children often develop symptoms of insecurity and fear of directly expressing their honest emotions and opinions. In addition, the traditional value system and patriarchal and authoritative social norms that shape adult-child relationships in the Arab society may also have contributed to this perception (Nasser, 2002).

Teachers agreed that the larger conflict between both people is a significant challenge for teachers and the school in general. It influences the school and constantly changes the internal dynamics. Recently, Palestinian teachers were more eager to talk about current events while Jewish teachers like to leave that outside the school curriculum. One Jewish teacher mentioned that she does not feel comfortable talking about her child's military experience because she does not want to offend Palestinian teachers. Jewish teachers think that they cannot make any complains about the other side because they will be accused of discriminating against them. Palestinian teachers think they work harder because they have to translate for their Jewish peers who do not master Arabic. This reality increased the level of tension and a strategy of avoidance has developed as the primary tool in dealing with the constant escalation outside of the school community. While Palestinian teachers had the urgent need to discuss events in classes, Jewish teachers felt both incapable and fearful that such conversations might "get out of hand."

In exploring children's perceptions of their school's strength, sixth and third grade students (13 and 16 respectively) valued the fact that they learn "together." Indeed such joint space is rare for Palestinians and Jews in Israel because those communities are mostly segregated, with the main interaction being through economic exchanges. It is important to highlight that the majority of the children who were interviewed lived outside of NS/WAS (9 lived in the community and 20 outside). Children also pointed out that they have more holidays and celebrations than regular schools. Students not only enjoy the various religious celebrations but they also believe they get more days off school because of that. These holidays vary from religious celebrations of the three religions represented in the school to national holidays and commemorations.

Jewish children were excited about learning two extra languages and Palestinians were excited about their fluency in Hebrew. Despite the fact that Jewish children do not have any fluency in Arabic, in their perception it is a positive thing to know the other side's language. Again when compared with regular schools where Arab children are obliged to study Hebrew starting in second grade, while Jewish children may elect to study Arabic in high school. Considering this outside reality, it seems that the bilingual nature of the school offers Palestinian children a more effective environment to practice their Hebrew and as a result they become more fluent, while Jewish children become familiar with the idea that learning Arabic is normal and necessary.

Most children interviewed liked their teachers and their different styles of teaching. Even though sixth graders had generally less favorable views of teachers than third graders they all enjoyed playing outside together and they especially enjoyed the swimming pool in the village. Palestinian children added they liked the school because of the freedom they enjoyed there, and they compared that school to their local schools that operate on a very strict and traditional curriculum. They also appreciated the school's emphasis on self-expression and raising confidence that suggests that teachers' attempts to empower Palestinian children have been effective.

Children pointed out that the tension that exists between children in the school is a challenge. Some of this is based on fights "between boys from both sides," as several children stated. Children also mentioned that they have fights over ethnic issues such as "Nakbeh day" and "Yom Zekaron." Overall, there seems to be a number of "behavioral issues" at the school that was indicated by interviews with teachers and parents. Several girls claimed that "Jewish boys have many misbehaviors." There was a general feeling among all children that the school had deteriorated in the past few years. Many especially girls in sixth grade expressed their discontent with the school's environment. One girl said, "The school used to be much better. Now, I am learning about Arab holidays more than I learn about mine." Another added, "We used to learn more together. Now we hardly do that." Obviously, disputes and formation of small groups are part of any school culture. However at NS/WAS the binational school interaction and the general violent conflict context could easily polarize the typical dynamic into an ethnic one.

The distance to school and from each other has contributed to the tension and has limited most Jewish and Palestinian from extending friendships beyond the school. When asked about friendships, the few visits that children talked about were of Palestinian kids visiting Jews and not the opposite (there were no visits of Jewish kids to Palestinian homes). It also seems that the few visits that exist are mainly between children who live in NS/WAS and others outside the village. The violent escalation of conflict during the "al Aqsa Intifada" has added tension and distrust between the Arabs and Jews in general and probably added to the complicated arrangements for play-dates after school. In addition, Jewish and Palestinian children had no contacts with the other side before coming to this school. Jewish children shared that they experience antagonism expressed toward them by other kids in their community because they attend this school. Furthermore, Jewish children in general did not like the fact that they were, in many cases, the only children attending the school from their towns. Thus an after school visit would jeopardize the child's existing social network in his/her own community. On the other hand, Palestinian children expressed that their friends and relatives envy them. Again establishing social relationships and expanding personal networks to include members of the Jewish majority can be viewed as an element or indicator of social mobility among certain minority members.

## A Curriculum in Bilingual Education: Successes and Challenges

Many of the issues raised by teachers, administrators, and children had to do with the bilingual program in the school. There is a high level of frustration and tension around this issue amongst all. For example, Palestinian teachers' major concern is in learning and teaching Arabic especially to Jewish children. They all expressed concerns regarding the nature of the bilingual program and the lack of emphasis on Arabic. "We failed in teaching Arabic to Jewish students and Palestinian kids gave up their language for Hebrew" was a statement that echoed in many interviews with Palestinian teachers. Due to the reality of minority-majority relations in Israel, minority teachers and students are more exposed to Hebrew and their economic and civic survival force them to master the language and learn it as early as possible. On the other hand, as members of the majority, Jewish teachers and students have limited access to the minority language and need not learn it to succeed in life. Several Jewish teachers also believed they do not need to know Arabic to succeed in teaching. Due to these asymmetric relations, members of the majority tend to be less motivated to learn or pursue the language of the minority. Such a tendency becomes more problematic in a bilingual school in which the primary premise of the school is an equal treatment of both national groups and in many cases there is a need to adjust the time allocated to the acquisition of the majority language to empower the minority language.

Israeli Jewish teachers' main concern was the school's recent emphasis on achievement and academics. In their perception NS/WAS primary school is unique in its informal and flexible environment in which special care and conditions were provided to all kids based on their needs. However, when some parents complained about the level of academic achievement in the sixth grade (prior to their transition to middle school outside the village), the administration began emphasizing the need to be more rigorous academically. Obviously this tension is not unique to this bilingual school; nevertheless it does illustrate the gap between the Israeli Jewish teachers' perceptions of the school and its mission and many of the Jewish parents who were not satisfied with the academic performance of their kids when compared to their peers.

This pressure in part spearheaded the decision to teach mostly in uninational groups in which children rarely have classes together. This separation of both sides in most of the subjects has affected other aspects of the curriculum such as current events and educational hours.[5] One should mention that all teachers complained about how demanding Jewish parents were. Many said they interfere "too much" in the school. Half the number of the Palestinian teachers thought that Jewish parents are suspicious of them. Teachers agreed that parents' overinvolvement is a source of pressure and tension in the school. Although this challenge is not unique to a bilingual school, the Palestinian-Jewish dynamics complicates the context and adds tension to teachers' relationships, especially if they have not developed a coherent educational philosophy on how to deal with parents and other challenges.

In order to examine children's perception of their secondary language skills, they were asked to rate their Arabic/Hebrew fluency from 1 (not fluent at all) to 5 (very fluent). Children's responses confirmed teachers' views about the imbalance between Arabic and Hebrew. According to table 8.1, Arab children perceived their knowledge of Hebrew as being higher than Jewish children's knowledge of Arabic. A greater difference emerged between Arab male and female children in their self-evaluation of knowing Hebrew. Such a difference can be attributed to gender since Arab boys and girls have different opportunities to interact with the other outside of the school and different patterns of socialization.

The dominance of Hebrew at school was confirmed in observations and interviews when Palestinian children indicated they use Hebrew when talking to Jews

**Table 8.1** Language Fluency

| Jewish children | Arabic knowledge by gender (5 highest– 1 lowest) | Palestinian children | Hebrew knowledge by gender (5 highest– 1 lowest) |
|---|---|---|---|
| Males (9) | 2.8 | Males (7) | 4.7 |
| Females (5) | 2.6 | Females (8) | 3.6 |

and Arabic when talking to Arabs. When asked to explain why they do not use Arabic with Jews they all answered "they won't understand us." Furthermore, Jewish children rated their knowledge of English as higher than their knowledge of Arabic. They also added that they hardly speak or use Arabic except in Arabic lessons. Due to the imbalance in power relations, especially the minority members' dependency on majority institutions, Jewish children do not necessarily develop the need to know Arabic. In addition, the social and political context pressures them against such internalization of the language. This asymmetric usage of both languages was on the mind of Palestinian children who complained about the dominance of Hebrew in school and the lack of time spent on Arabic. One child said in the interview: "If you're good in math and science you have to learn in Hebrew and not Arabic." The general impression is that the separation of children contributed to the decrease in interest to learn Arabic among Jewish students. The school today reflects bilingual Palestinian students but the same cannot be said about Jewish students. Parents' surveys have supported this conclusion. When they were asked about reasons to choose this particular school, Hebrew was the first reason to choose the school among Palestinians and learning Arabic was not mentioned at all by Jewish parents. A lack of clarity about their bilingual goals and philosophy from teachers and administration made it easier for parents (both Jewish and Arab) to pressure the school in emphasizing content knowledge in Hebrew.

### Views about the Conflict

A bilingual training does not necessarily prepare teachers to handle conflict issues. In general, teachers interviewed believed they lacked the proper skills to deal with certain events commemorated at the school such as the "Nakbeh Day"[6] and the Day of Independence. They also shared that they received peace education training only once in the form of a two-day workshop a few years previously. Teachers expressed their need for further training to prepare them to handle current issues. For example, during one of the observations, a fight started between children on the day of commemorating the "Nakbeh" and the two directors' interference was required immediately to comfort everyone involved. In order to enhance the schools' ability to address these dynamics, such historical and symbolic celebrations can be planned and carried out by staff members who have proper skills in the area of conflict management and resolution. This is especially important at times when teachers themselves are experiencing mixed feelings and despair about the chances for dialogue between both people.

Obviously having a bilingual and binational school in a conflict setting like Israel and Palestine is highly different from other areas. Conducting bilingual education in such an intense conflict context imposes certain challenges and necessitates teachers and staff to acquire a specific set of skills. In NS/WAS, especially among teachers, the general impression is that the school's founders' mission and vision to educate for peace and tolerance between both people has been challenged by the Intifada events. In the past three years and in the midst of changes in administration and a rapid expansion, teachers voiced concern that the school has deteriorated in its accomplishment of bilingual education. Teachers and staff indicated that the primary school was more successful in accomplishing its goals when the school was smaller and contained students mainly from the village.

Still, in the midst of this critical atmosphere, something positive was happening at the school that indicates a degree of success in achieving the school's mission in peace education. The school's ability to provide a meeting space for both people to learn and work together is an important step in educating for peace and in creating change in attitudes and stereotypes. Children's high level of understanding of both sides of the conflict and possible future resolutions provides evidence of that. Children as young as third grade expressed a desire to share the land and solve the problems peacefully between both people. Children, overall, recognized a special role that they might play in the future to solve the conflict between Palestinians and Israelis but they could not pinpoint or specifically describe the nature of that role. Some children from both sides said that the two people should "talk instead of shooting at each other and to seek peace and compromise." Also few others said that they "want to live together without the wall that's being built between them." The binational school reality in addition to the emphases on peace education and the school's mission has contributed to such perceptions among children.

To get more specific insights into the schools peace education input, children were also asked to recognize pictures of leaders from both sides and share what they think about them. Most children recognized the figures we shared with them, for example pictures of people praying from both sides and political leaders. Few had difficulty identifying pictures of Sharon and Arafat. All Palestinian children expressed "dislike" of Sharon and characterized the conflict between both peoples as caused by disputes over land. Jewish children, on the other hand, characterized the conflict as being more cultural (misunderstanding/ lack of knowledge) and personal between the two leaders.

Children's responses in this category indicate a high level of awareness regarding the political context because they were able to identify political leaders and express certain views with a degree of confidence and ability to articulate an opinion. In addition, they were aware of the conflict and some of its major issues (land, occupation, political leaders, the wall, etc).

Considering the age and context of the Arab Israeli conflict and despite the challenges facing the school, it is clear that both Arab and Jewish children welcome the opportunity to interact and learn with each other in a more pluralistic environment than other schools. For example, none of the children interviewed expressed any prejudice or categorical level of animosity toward the opposite national group. In a regular Palestinian or Jewish school, expressing negative stereotypes and prejudice against the other is an integral part of these children's reality (Abu-Nimer, 2000).

In conclusion, NS/WAS children's perceptions of the conflict and their resolutions reflect a high degree of awareness. The children expressed no fear of interacting with peers of same age and teachers from the other nationality, and also suggested steps to resolve the conflict while taking into consideration the two national groups and their aspirations. Such responses suggests that the "school's peace education ideology" and message are reaching the students to varying degrees.

## Bilingual Education as a Means for Peace Education: Lessons Learned

Teaching the other's language is not enough to promote peace education especially in a context where the majority is in conflict with the minority group. In such contexts a deliberate peace education orientation is a necessary component of bilingual and binational education system, and could be an effective tool in creating a culture of peace. In Israel-Palestine, the bilingual and binational educational settings without peace education orientation (without tools and skills to handle their perceptional differences) can result in negative perceptions and attitudes about the other. Thus, any bilingual schools' mission has to be constructed within a peace education paradigm.

Furthermore the provision of a joint space for bilingual and binational education is not enough to ensure coexistence and understanding of the other. Most importantly, it is not enough to create a condition for equal status of both people and their languages. Empowering the minority group language should become a priority in these types of schools to create equality. Ongoing teacher training and awareness workshops should take place, planning creative forums to empower minority students and teachers and to express views about current issues is important too. Also, involving and educating parents in the process are possible arrangements to handle the effect of the conflict on the bilingual and binational school.

There are certain principles that can contribute to the success of a bilingual school in Israel. Based on the NS/WAS school experience, one can identify several of these. Constructing an educational environment facilitated by these principles can be a model for any bilingual program in areas of conflict. The following are some of these principles based on the findings.

- Institutionlizing Equality between Arabs and Jews
  Institutionlizing equality between Arabs and Jews "on all levels in the school," as stated by NS/WAS school's mission, is a crucial condition for success of any joint educational institution in Israel. This type of equality can be reflected in staff, directorship, teachers, publications, and all other aspects of binational and integrated schools. Considering the fact that as a bilingual and binational school that operates in a context of political violence and deep rooted system of dominance of Jewish majority over the Palestinian minority, NS/WAS has been successful to a degree in implementing this condition of equality in distribution of resources. The school's emphasis on having two codirectors, equal number of teachers, and equal amount of time for cultural and religious celebrations supports this principle. In addition, attending to the cultural and national needs of the Palestinian children rather than imposing the Jewish Israeli majority beliefs and narrative certainly reflects the school's ability to address the different needs of the minority and majority on that dimension.

- Promotion of the minority language
  For any bilingual school in Israel it is essential to ensure that the students of the majority group are sufficiently exposed and master the minority language. Thus Jewish students in bilingual schools need support and an environment that encourages them to become fluent in Arabic like Arab students who earn Hebrew. Providing such an environment in a reality in which the majority has no cultural, economic, or social need for the Arabic language is the most challenging aspect for the school. As one teacher who is a previous administrator explained, "We have to aim for a 70 percent Arabic and 30 percent Hebrew model to succeed. The problem is how to convince Jewish parents that this will not harm their children's academic potentials." Constructing a policy that includes hiring teachers who speak the two languages, converting the curricula into a complete bilingual program, opening courses for parents to study both languages, using extracurricula activities to deepen the learning of foreign language, and many other initiatives should be taken.

  The dominance of Hebrew detracts from the school's ability to offer a comprehensive bilingual education to its maximum potential. However, since there is no standard curricula in bilingual education in Israel, and due to the school's status in the ministry of education and its history, NS/WAS primary school does not have a set of guidelines on how to systematically introduce bilingual education in all aspects of its operation. Such guidelines and standards are necessary for any bilingual school in Israel.[7]

- Planning for intercultural dialogue
  Binational schools in conflict areas are a perfect set up to construct educational conditions for an environment of an ongoing dialogue between both people. The students are meeting and in contact with each other on daily basis. Instead of relying on such a fact to produce understanding and sense of positive coexistence, the school ought to be more proactive and design a systematic plan to provide an intentional experience of dialogical encounter as opposed to random and unorganized "contact."

  Thus the school has to provide certain conditions to make the contact successful in producing positive attitudinal changes. It is imperative that integrated schools take such issue seriously due to the risk that their school can become a space for negative contact, especially with the outside contextual factors creeping into the school dynamics.

  To convert the school to an ongoing dialogical encounter, staff (administrators and teachers) has to be trained on regular basis as facilitators in intergroup relations. Furthermore a special school time should be allocated to dialogical conversations as part of the formal curricula (such a school should not leave the interethnic relations outside of its gates and pretend that the conflict does not exist and create an image of harmony and peace within its walls).

- Learning about other cultures
  Bilingual schools can be an ideal space for learning the culture and tradition of both people in a safe environment (for example, through celebrations and holidays). The NS/WAS school emphasized through the many religious

celebrations and holidays that the school has an intentional and direct policy to help students learn each others' cultures. In fact, some believe it has been "overemphasized" to a degree that it is taking a lot of instruction time from children. However, school officials have to resolve controversies around these celebrations and be able to design a curriculum to guide their teachers on how to conduct such celebrations.

- Conflicting priorities
  In integrated bilingual schools there is a major challenge to balance between the traditional academic requirements and the flexibility and creativity required in dealing with two or more national groups' narrative and identity. Thus, nontraditional and more alternative pedagogy are almost a must for the school to be able to address its complex dynamics. In the case of NS/WAS, the school was under a lot of pressure from the Ministry of Education and parents to focus on academic performance and knowledge. In many ways, this restricted the teachers' use of alternative and nontraditional methods of teaching. Parents' impression is that teachers are not teaching especially when they take a walk or work on a project outdoors.

In conclusion, the principles mentioned are conditions to empower the academic program at the school as well as the impact of the contact provided for both people. Contact alone is not enough to bring change on the macro and structural levels (Abu-Nimer, 1999) even though it has provided opportunities for change on the micro/individual levels. The results of this study of teachers and students views about the school, the other side, and the conflict in general should spark more interest in these initiatives and their contributions to a much larger change in Israel and other conflict areas. They will also raise the awareness of Israeli educators to the need for teacher preparation programs that address peace education and bilingualism in Palestinian and Jewish Israeli schools.

Finally, NS/WAS has its unique principles and status as the only joint living space in the country. The primary school can serve as an advocate for equal educational opportunities and equal investments in both languages. More dialogue among the four bilingual schools on pedagogy, curriculum, successes, and challenges would contribute to developing the philosophies and mission of these schools. It would also help create a much needed support system for teachers and administrators involved in bilingual education in the country.

## Notes

1. See special issue of JPD (2005), 2 (2).
2. For further discussion of these policies see: Abu-Nimer, 1999; Jabareen, 2005; Rouhana, 1997.
3. See NS/WAS website for more information about this school.
4. The two authors with two field researchers (a Jewish and a Palestinian) spent many hours interviewing in depth and meeting with the principles, all teachers, and students. The interviews included previous Palestinian and Jewish administrators. Interviews were also conducted with graduates and parents but are not reported in this chapter. Classroom observations and school documents were also used to gather information.

5. The rational for separation was mainly to enhance the academic performance of the Arab students in Arabic, math, and sciences, and due to the fact that most of the Jewish teachers did not master Arabic.
6. Nakbeh day commemorates the expulsion of Palestinians out of Palestine in May 1948.
7. It should be noted that NS/WAS school has experimented with different models of bilingual teaching. The data for this chapter was drawn from a period in which the school has decided to conduct several classes in a uninational format (for example math) also in classes where Jewish teachers did not know Arabic and there was no Arab coteacher with them.

## References

Abu-Nimer, M. (2000). Peace-building in postsettlement: Challenges for Israeli and Palestinian peace education. *Peace and Conflict: Journal of Peace Psychology*, 6(1): 1–21.
Abu-Nimer, M. (1999). *Dialogue, conflict resolution, and change: Arab-Jewish encounters in Israel*. Albany: SUNY.
Allport, G.W. (1954). *The nature of prejudice*. Reading, MA: Addison-Wesley.
Bekerman, Z. (2003). Reshaping conflict through school ceremonial events in Israeli Palestinian-Jewish co-education. *Anthropology & Education Qaurterly*, 34 (2): 205–224.
———. (2004). Multicultural approaches and options in conflict ridden areas: Bilingual Palestinian-Jewish education in Israel. *Teachers College Record*, 106 (3): 574–610.
Cook, S.W. (1978). Interpersonal and attitudinal outcomes in cooperating interracial groups. *Journal of Research and Development in Education*, 12: 97–113.
Feuerverger, G. (2001). *Oasis of dreams: Teaching and learning peace in a Jewish-Palestinian village in Israel*. New York: Routledge.
Garcia, O. (1997). Bilingual education. In F. Coulmas (Ed.), *The handbook of sociolinguistics* (pp. 405–420). Malden, MA: Blackwell.
Glazier, J. (2004). Collaborating with the "other": Arab and Jewish teachers teaching in each other's company. *Teachers College Record*, 106 (3): 611–633.
Hewstone, M. and Brown, R. (1986). *Control and conflict in intergroup encounters*. Oxford: Basil Blackwell.
Nasser, I. (2002, Fall). The ideas of parents and children about the importance of developmental skills among Palestinians in Israel. *International Journal of Early Years Education*, 10(3): 215–225.
Pettigrew, T. (1986). The intergroup contact hypothesis reconsidered. In M. Hewstone and R. Brown (Eds.), *Control and conflict in intergroup encounters* (pp. 169–195). Oxford: Basil Blackwell.

# CHAPTER NINE
# Is the Policy Sufficient? An Exploration of Integrated Education in Northern Ireland and Bilingual/Binational Education in Israel

*Joanne Hughes and Caitlin Donnelly*

### Introduction

For the past few decades contact interventions have played an important role in efforts to manage conflict and promote better relations between Protestants and Catholics in Northern Ireland (Shared Future Document, 2003) and Arabs and Jews in Israel (Bar-Tal, 2000). The target for many such initiatives has been children and young people. Reflecting on this and in parallel developments, it is found that both countries have seen the creation of new and innovative approaches to educating, together, children of diverse religious, political, and national backgrounds. In Northern Ireland integrated schools emerged in the 1980s as an alternative to existent segregated schools. At around the same time, bilingual, binational schools were first established in Israel as an experiment in Arab/Jewish coeducation. In both contexts the schools reflect an ideological commitment to a more peaceful society and are based on the theoretical premises of the contact hypothesis (Allport, 1954). Based on qualitative research in four schools, this chapter examines the nature of the contact experience in two integrated schools in Northern Ireland and two bilingual/binational schools in Israel. Through comparative analysis, it illuminates some of the contextual and process variables that seemingly mediate the quality and moderate the effectiveness of contact in each school setting. The final section highlights some issues that may be of relevance to policymakers and practitioners.

### Mixed Religion/Binational Education in Northern Ireland and Israel

The first binational, bilingual school was set up in Israel in 1984 but did not receive government support until 1993. Similarly, planned integrated schools (or Grant Maintained Integrated schools) in Northern Ireland were established in 1981 and received full government recognition and financial support only by the terms of the 1989 Education Reform (Northern Ireland) Order. Transformed schools also exist in both countries, whereby existing segregated schools have opted to transform to integrated status.[1] The clear similarities of principle that underpin integrated and bilingual/binational school types are summarised below:

- Participation of all religions/cultures
- Provision of a natural ongoing framework that enables the day-to-day meeting between children of the two cultures
- Nurturing each child's identity by imparting knowledge of his/her culture and tradition while inculcating knowledge and respect for the culture and tradition of others
- Searching for common ground between the two cultures.

Yet there are differences in the systems too. There is a clear bilingual focus in the Israeli system where the aim is for Jews and Arabs to learn both Hebrew and Arabic as one way of raising intercultural awareness. Bilingualism is not a feature of integrated education in Northern Ireland where the dominant language for both Catholics and Protestants is English. In addition, the acceptable parameters for balance are more rigidly defined in Israel where Arab/Jewish staff and student ratios can deviate only marginally from 50:50. In Northern Ireland the limits of Protestant/Catholic balance are more flexible. The Department of Education accepts a ratio of 70:30 for planned integrated schools (with 30 percent of pupil intake representing whichever is the minority community in the area where the school is located). Transforming schools can attain integrated status if they can prove that a minimum 10 percent of incoming enrolment is from the minority community in the school's catchment area. Such schools are expected to achieve a 70:30 balance within 10 years (Department of Education (Northern Ireland), 2004). In both Northern Ireland school types it is expected that schools will aim toward a 40:40 representation of teachers and governors from each religious group.[2] But perhaps the most important difference is the distinct raison d'être embraced by each of the school types. Whilst schools in both contexts represent an alternative to segregated education systems, Israeli schools are variously designated bicultural, binational, or bilingual, portraying an orientation that is concerned with promoting cultural coexistence in a divided society. Schools in Northern Ireland are referred to as integrated, and whilst proponents of the schools would contend that this does not refer to any pancultural intent (NICIE, 2004), the emphasis in the Northern Ireland integrated schools is firmly on the promotion of similarity between the two religious groups. Despite these ideological differences, fundamental to both school types and setting them apart from the norm in both jurisdictions is the opportunity they present for sustained contact between members of ethnically divided groups.

## A Conceptual Framework

Understanding how contact serves to promote attitudinal change has been the subject of sustained theoretical research in recent years. Pettigrew (1998) identified four interrelated processes through which contact can mediate positive attitudinal change. These include: (1) *learning about the out-group*, where optimal intergroup contact generates new learning that helps correct negative views of the out-group; (2) *changing behavior*, where optimal intergroup contact acts as a benign form of behaviour modification that can lead to positive attitudinal change; (3) *generating affective ties*, where positive emotions aroused by optimal contact can also mediate intergroup contact effects by, for example, reducing anxiety and generating intergroup empathy; (4) *in-group reappraisal*, where optimal intergroup contact provides insights about the in-group as well as the out-group that challenge the taken for granted in-group norms and customs as the only way of effectively managing the social world.

In a meta-analysis, Pettigrew and Tropp (2006) also highlight negative process issues that can militate against positive attitudinal change and reinforce or heighten prejudice. These relate primarily to the sense of threat and anxiety experienced by participants in contact encounters. Pettigrew and Tropp report that anxiety is a much more significant mediator of intergroup contact reducing prejudice than increased knowledge of the out-group. Similarly, liking the out-group can lead to significant shifts in affective response even when stereotypes remain largely unchallenged by the production of new information.

Complementing reviews that have focussed on *mediators* of contact, research has also accumulated evidence that group salience (i.e., the extent to which *group* memberships are psychologically "present" during contact) is a key *moderator* of the effects of intergroup contact on prejudice reduction (Brown and Hewstone [2005]; Hewstone, 1996; Hewstone, Rubin, and Willis, 2002; Tausch, Kenworthy, and Hewstone [2005]). Studies show that the generalization process, from the judgments concerning single individuals to the whole out-group, is favored by the presence of a link between these individuals and the group. When group salience is low, the situation is interpersonal with *individuals* responding to other individuals on the basis of idiosyncratic characteristics (that might include, for example, shared interest in sport, drama, music, etc.) rather than group affiliations or identities (e.g., Arab, Jew, Protestant, Catholic). Where contact is interpersonal, more positive responses to out-group members are not likely to extend beyond *individuals* to incorporate the group as a whole. One way of increasing the potential for generalization is to make group categories salient during contact.

Focusing on mediational and moderational factors, we present data derived from an exploratory study, undertaken during 2003, of contact in integrated and bilingual/binational schools in Northern Ireland and Israel.

## The Research

The research approach employed here is in the qualitative, naturalistic tradition of Geertz's "thick description," where emphasis is on achieving a "depth" of understanding through detailed and nuanced description of the issues under study (Geertz, 1973). The aim is to explicate contact issues, *in context*; therefore no attempt is made to extrapolate beyond the immediate research setting and no claim is made as to the generalizability of our findings. Data collection was primarily through in-depth semistructured interviews (30) with school principal(s), teachers, and parents in each of the Northern Ireland and Israeli schools. Interviews were complemented by a document review of relevant literature (prospectuses, minutes of governors' meetings and annual reports), classroom observation in Israel, and informal discussions with teachers and children in each of the four schools during a series of visits. The interviews lasted, on average, an hour and covered issues such as understanding of, and value attached to, the integrated/bilingual/binational approach to education in a divided society; understanding of, and commitment to, the delivery of the school's core values (tolerance and respect); perception of facilitating/inhibiting factors in relation to the delivery of those values; the practices adopted, challenges faced, and mechanisms developed to overcome perceived difficulties in achieving them and perceived effectiveness of schools in meeting their stated objectives and having an impact on negative social attitudes.

## The Schools

The four schools selected for this study are at primary level and have a pupil age range of 4–11 years. One of the Northern Ireland schools (School A) is planned integrated, and was established in 1987, the other (School B) is transformed integrated. Both Israeli schools are planned integrated, one (School C) was established in 1984 and is located in a rural Arab/Jewish "purpose built" community, the other (School D) was founded in 1998 and is situated in the center of a large urban town. Three of the schools are of similar size with student numbers of around 200. School B, with a pupil enrolment of 111 is smaller than the others. Staff/student ratios vary significantly between Northern Ireland and Israel with the Israeli schools having almost twice as many teachers per student as the Northern Ireland schools.

## Findings

### Creating a "balanced" environment

In both Israel and Northern Ireland the pursuit of *balance* is indicated as the means through which equal recognition is accorded to each religious/national group. The interpretation of the term is though different in each context. In Northern Ireland, balance is considered a defining *condition* of integrated education and is used as a physical reference to pupil intake and staff/board of governor recruitment patterns that *aim* toward numerical equivalence of Protestants and Catholics. In Israel, balance is presented as a more all-embracing, pervasive concept, relating not only to a numerical balance of Arab/Jewish pupils, but also to *processes* through which a sense of equality and mutual understanding, in the school context at least, can be realized. These processes are given expression in "perspective-taking" and "bilingualism."

### Perspective Taking

There is broad consensus in both Israeli schools that children should be exposed to the historical and political narrative of each national groups. In order to achieve this, significant time is given over to the preparations and delivery of classes/sessions that tackle issues of division and to discussion about which Jewish/Arab national days/holidays should be celebrated, including the format of such celebrations. In School D, in particular, "partner" teachers[3] indicate that working together to prepare classes often helps each teacher appreciate and better understand "out-group" perspectives that deviate from their own, thus enabling them to be more "informed" and "sensitive" in their communication of Arab/Jewish relations to children. In addition, working through sensitive and difficult issues often demands significant self-disclosure and can lead to the development of deep affective ties between teachers that sometimes mediates an approach to the delivery of information in the classroom context that is emotionally charged. The following example serves to illustrate this point.

> Children like to listen to our stories and like to tell their stories. When the teacher tells them, "This happened to me but look at me and where I am and I want peace," then they [the children] talk about it . . . Mari the Jewish teacher who is with me . . . she has a background with all that stuff and we both decided that [to talk openly about our personal experiences] is a solution. On [Jewish] Independence Day we talked about it . . . As fourth graders they might not understand everything, but they tell their stories and we tell them our stories and the teachers talk together and then we go into the class rooms and talk to the children. I can tell them that this is the first year that we had Nakbe together and that is the day that the Palestinians lost their land, and another teacher [the "partner" Jewish teacher] came in and we started talking and then I started crying and she said, "I will help you" and I had to leave the class and the Jewish children were wondering why I was crying and [when I came back] I told them the story [of my experience] and they felt so close to the background of the Palestinians. I think that the teacher has a big role . . . at the end of the lesson all the girls came and started hugging me. [Arab teacher, School C]

Clearly, the personalized and emotional communication of what Nakbe Day meant for this Arab teacher, the sensitivity displayed by her Jewish colleague, and the opportunity for children to explore alternative perspectives facilitated a situation wherein Jewish children could understand and empathize in a very direct way with the experience of Arabs. Pettigrew (1997, 1998) argues that drawing on personal experience (self-disclosure) is a central process in cross-group friendship. By self-disclosing, disclosers tell others how to understand the way they see themselves, or how to empathize with them. Central to the notion of self-disclosure as a mediator is the idea that it establishes mutual trust and detailed knowledge about the other party that may disconfirm stereotypes.

Not all teachers in the Israel schools are as comfortable dealing so openly and intimately with potentially volatile, divisive issues. In School C, where there is a less collegiate approach and where only some teachers have a "partner" arrangement with a teacher from the "out-group," the preference articulated by most is for a much less personalized approach. Values such as tolerance, respect, love, and forgiveness are discussed with children, but only in the absence of any direct reference to Arab/Jewish relations. Some teachers observed that this approach was less than satisfactory as pupils became confused at the contradictory nature of information presented to them. On the one hand, they were being encouraged to show love and respect for everyone. On the other, the actions of the out-group toward them and their community (interpreted within one's own group frame of reference) were conceived, at best, as hostile and negative. Unable to appreciate the worldview of "the other," children struggle to relate the values they are being persuaded to espouse to the out-group.

## Bilingualism

The relationship between the bilingual approach to education, which is a feature of both Israeli schools, and the promotion of equality, tolerance, and respect in the school context can be complex and is influenced by the status of Arabs and Jews in wider society. Protagonists of bilingualism believe that it is an important facilitator of equality because the status accorded to both languages places Arabs, who are the minority, subordinate group in the wider Israeli society, in a position of strength. This is because Arabs are more likely than Jews to be bilingual on entering the school as a consequence of living in a Jewish state where they are more frequently exposed to Hebrew (which is the official language) than Jews are to Arabic. The greater competence of Arabs in Hebrew, vis-à-vis Jewish lack of competence in Arabic, places them (Arabs) in a more dominant position in an area of school life that is of paramount importance and is seen as "correcting" a power imbalance that prevails in wider society. Some teachers indicated that Arabs having better language skills than their Jewish counterparts is important for the development of good relations between the two groups. At one level, it helps generate in Arabs a sense of self-worth and confidence that can ameliorate some of the fear that often accompanies interaction with Jews. At another level, it can help dispel some of the stereotypes that Jews as the dominant group might have of Arabs regarding inferior intelligence and ability. Moreover, it is hoped that the cooperation between Arabs and Jews, necessitated by the bilingual approach to learning, will help promote intergroup respect.

Juxtaposed with the positive aspects of bilingualism, there is also a perceptible tension between embracing aspects of out-group culture in the pursuit of improving relations and more self-interested objectives in an unequal society. This point is illustrated by a debate sparked in School D relating to the abandonment of the use of both languages outside of the classroom in favor of communication in the more common language of Hebrew. Jewish teaching staff were concerned that this practice undermined the value of the Arabic language and, by definition, Arab culture and identity. When, however, the parents of Arab children were consulted about the issue, they were seemingly unconcerned. Parents placed greater priority on their children becoming competent in the language of the majority than on the promotion of what some saw as "artificial" equality through bilingualism. Unlike Jewish parents, who tend to send their children to the bilingual/binational schools out of an ideological commitment to the aims of the school, Arab parents tend to be more pragmatic. Many Arab children attend school because their parents believe it represents the best opportunity for them to get an education that will equip them with the necessary qualifications and language skills for securing a good job in Israel. This example clearly indicates the wider societal dynamic, confirming Pettigrew and Tropp's (in press) observation that majority group interpretation of equality can be perceived by the minority group as "patronizingly unequal," and supporting a point made by Bekerman (2002) that wider national disputes often inform the nature of contact initiatives.

Where tensions do arise, an interesting question is whether or not they impede efforts to improve intergroup relations. Our data suggest that an important issue in the resolution of conflict or disputes that arise in the schools is the willingness of facilitators to explore differences that may have caused the problem in the first instance, and this correlated with an overall sense of purpose.

## Commitment to Purpose

In Israel commitment to "relationship building" is a priority and pervades all aspects of the school structure and the content of curriculum delivery. Teachers, in consultation with each other and with parents, are constantly seeking the "best ways" of meeting school objectives, whilst at the same time being sensitive to the different cultural and political positions of the two groups. A very different situation pertains in Northern Ireland where teachers and parents in the integrated schools are reluctant to deal with divisive issues, either amongst themselves, or in their relations with children. Group issues are played down, to the extent that the use of the terms "Protestant" and "Catholic" is considered somewhat taboo because it could draw attention to differences that the schools actively seek to suppress. In one school this even extends to expressions of cultural identity, through sport.

> There will always be issues that will never be resolved ... Like I don't know what it would be like here if they were to introduce Gaelic football ... like no one has brought that up, they [the Board of Governors] are very neutral now ... I don't know how it would be if they were to say, "we are going to bring Gaelic things in" ... I don't know how that would happen as that would be seen as being reflective of one side or the other. (Parent, School A)

Gaelic football is considered a completely benign aspect of cultural identity in Catholic schools. In Northern Ireland, however, anxiety about highlighting any aspect of group distinctiveness relates to a shared cultural norm of avoiding issues that might potentially offend the "other" community, sport included (see, for example, Blacking, Holy, and Stuchlik, 1977; Buckley, 1989; Donnan and McFarlane, 1983; Harris, 1973; Hughes, 1992; Larsen, 1982; Leyton, 1974).

Related perhaps to a fear of offending, many in the Northern Ireland schools believe that the role of the school should not extend beyond the provision of a shared "neutral" space within which Protestant and Catholic children can be educated together. Emphasis is placed on the promotion of a common "school" identity, with tolerance and mutual understanding seen as "natural" and unquestioned outcomes of contact that may elude children in other areas of their lives. Success is measured in terms of not "rocking the boat" with little attention paid to how, or even if, contact is effective.

> The kids come to school and they can play together and work together and live together from 9 to 3 so why fix it. We don't set out to teach tolerance but that is what comes out of it ... you tolerate the other person in the nicest possible way and we are friendly and warm towards one another. As long as what we do in school doesn't stir children then it is not a problem. If we are all just seen as the same and we all wear the [names School B] sweatshirt and we are proud to be [names School B] children rather than a Celtic supporter who comes to [names School B] or a Rangers supporter who comes to [names School B] and we don't explore those issues ... the general feeling is that we don't have to ... I think that what happens at staff level happens at class level and pupil level, in that we neutralize our conversation ... we neutralize everything so that we are all the same. (Principal, School B)

Whilst the emphasis on neutrality pervades relationships at all levels, there are clear examples in one school of intolerant and provocative behavior amongst pupils.

One teacher referred, for example, to the fact that some pupils come to school wearing Rangers and Celtic t-shirts under their uniform.[4]

> You know that underneath their uniform they [pupils] are wearing their Rangers shirts or Celtic shirts . . . I mean at the minute we are quite happy if they wear it underneath . . . because it is not offensive if it is hidden . . . openly[wearing such shirts] it is banned because there is a fear that it would stir trouble. (Principal, School B)

This example highlights the complex nature of intergroup relations in the school. The pupils clearly know they are engaging in offensive behavior (otherwise why bother wearing a football shirt *under* their uniforms) and the teachers clearly know they are doing it (because they mentioned it); yet such behavior is not categorized as problematical because, at some superficial level, pupils coexist without apparent antagonism. Even when sectarianism becomes more overt, the response of teachers is often evasive. Sectarian incidents are categorized by many teachers under the rubric of bullying. By focusing on the power relationship between the bully and the victim, the need to address the content issues of any dispute is circumvented. This lack of openness is pervasive in the school and has helped generate an environment of mutual suspicion, not just amongst pupils but also between Protestant and Catholic staff, to the extent that some teachers prefer to take lunch-breaks only with coreligionist colleagues. One longer-serving teacher observed that the exponential increase in integrated schools has to some extent been at the cost of the original vision.

> I think that the founders [of integrated education] were very sincere people who had a vision . . . You know it is like knitting a jumper and you follow the pattern, as you do more and more, then you get a bit more confidence and you follow the pattern a wee bit less but as you move away from the pattern the jumper doesn't resemble the pattern which is quite sad because really that is the jumper that you wanted to knit in the first place. (Teacher, School A).

The unwillingness or inability to tackle divisive issues in the Northern Ireland schools contrasts sharply with the Israeli schools where there is a common belief that *only* by focusing on the distinctiveness of the two national identities and the differences between them are the aims of the school achievable. The remainder of this section focuses on how *intergroup* contact is mediated in Israel.

### Anxiety Attenuating Mechanisms

Addressing issues of identity and difference in Israeli schools, particularly during times when conflict in wider society is raw and widespread, demands great sensitivity. The potential for damaging already fragile intergroup relations is a key concern of staff in both schools. One of the mechanisms developed to limit anxiety and the sense of threat that often accompanies intergroup work is a parallel intragroup/inter group approach to encounters that deal with sensitive issues. Well documented in community relations literature on Northern Ireland (see Hughes 2003), though receiving little attention in the theoretical literature on contact, the emphasis is on separating groups according to the religious/national identity characteristics that divide them for alternate intragroup/ intergroup limited periods. The rationale is that negative emotions can become heightened during intergroup situations, thus, providing an opportunity for individuals to vent their concerns in the "nonthreatening," intragroup context can act as a safety valve for ameliorating tensions that may have arisen during the intergroup session. In addition, the intragroup context can provide assurance for individuals that any reappraisal of own group values or "position" on the basis of new information communicated during intergroup sessions is supported or at least tolerated by the in-group. The outworking of the parallel intragroup/intergroup approach is described in the following example that refers to the celebration of Nakbe/Independence Day in School D:

> We decided that on the day we would have two hours together—binational—talking about independence and the pain and the price of the creation of independence and why the Palestinians didn't like this and why they fight. Then we separated and the Jews had their ritual for Independence Day and Arab children went to their class and talked about what it means to be independent, where does independence come from and why Nakbe is the opposite of independence. Then we talked about what it is to be Palestinian . . . and the Jews were talking about how they felt about these things. After this they came together and were mixed and divided into two groups [each comprising Arabs and Jews] each did something together. I mean one group did something on how can we share the country and two nations. One group did something about love together . . . each class did this. (Arab Principal, School D)

The intragroup/intergroup approach is also employed by teachers in School D who, each week, meet separately in Arab and Jewish groups to plan how aspects of the curriculum relating to cultural and national identity might be delivered; the two groups then come together to share ideas and discuss any points of contention. In addition, "Away Days" are organized so that teachers can dedicate time to reflecting on perceived good and bad practice during the previous teaching period. Such events also provide an opportunity to explore "cultural misunderstandings" and to deal with relationship issues/concerns that may have arisen between partner colleagues.

Given the centrality accorded to negative effect as an inhibitor of intergroup contact in contact literature, attenuating processes that can minimize anxiety and fear are clearly important. Their relevance is demonstrated when the Northern Ireland and Israeli data are compared in relation to perceived outcomes of the school experience.

As noted, the theoretical literature on contact makes the case that the effectiveness of contact is moderated by the extent to which group categories are salient during contact. In line with this, there is evidence, albeit, self-reported, that attention paid to promoting understanding and tolerance of difference that is so apparent in the Israeli schools does have important consequences for both teachers and pupils in terms of own group reappraisal and generalization.

All of those interviewed were convinced of the value of the bilingual/binational schools both in terms of promoting better relations between Arabs and Jews in the schools and in terms of potential spin-off effects in wider society where prejudiced and negative social attitudes had been challenged. It is interesting to compare the responses of Northern Ireland teachers and parents to similar questions regarding the effectiveness of integrated schools. Here, where both schools actively seek to decategorize the contact situation, "Flags and emblems divide and they mark territory and categorize so it is easier if we are all not categorized . . . if we are all just seen as the same" (Principal, School A), views are much less positive with some teachers actually questioning the value of integrated schools where the *integrated* dimension is so clearly absent. Moreover, parents and teachers indicate that any positive outcomes quickly dissipate when children leave primary school to attend segregated secondary schools.

## Discussion and Conclusions

The comparison between the Northern Ireland schools, where reference to difference is actively avoided, and Israeli schools, where significant emphasis is placed on group distinctiveness, suggests some of the processes through which genuine *intergroup* contact may be generated. Evidence from the Israeli schools lends support to Pettigrew's analysis that positive mediators include learning about the out-group; changing behavior; generating affective ties; and in-group reappraisal. Although qualitative in nature and therefore not sustained by empirical measurement, the Israeli data indicate that perspective taking and the bilingual approach to education may facilitate new learning about the out-group that can challenge previously held negative stereotypes and prejudiced attitudes. It would also seem that the means through which new information is communicated is important, thus personalized accounts and reference to the wider conflict context are seemingly more efficacious than abstract appeals to tolerance and respect.

Central to the potential for delivery of information relating directly to divisive issues between Arabs and Jews in Israel seems to be the strength of affective ties amongst teaching staff. Where relationships of friendship and trust have developed the exploration of contentious intergroup issues becomes a less threatening possibility, more intimate relationships, it seems, facilitate self-disclosure on conflict/divisive issues that, in turn, can lead to empathy and understanding of the out-group. In some cases the deeper sense of affinity with the out-group has led to a reappraisal of in-group position and to "new ways of thinking," about Arab/Jewish relations.

Although contact literature places emphasis on the importance of friendship, both in reducing the negative affect of contact and augmenting the positive affect (Pettigrew, 1997), there is little on *how* such friendships evolve. The research presented here suggests that in the Israeli context important predictors of friendship or, at the very least, supportive working relations are shared commitment to purpose, common aspirations/values, and mutual interdependence in delivering school objectives. It would seem that where facilitators agree on contact as a means of challenging the norms of existing intergroup relations, then a more collegial, proactive, and interventionist approach is the likely modus operandi. This has important implications for how potentially negative mediators of contact (anxiety and threat) are dealt with.

In the Northern Ireland schools, fear of repercussion informs a tendency to underplay differences between Protestants and Catholics and to avoid reference to any aspect of cultural/political identity that could be interpreted as contentious. In the Israeli schools too there is some anxiety that tackling divisive issues will exacerbate problems that exist between Arabs and Jews. The difference between the two school types in how negative affective response is dealt with correlates strongly with willingness to challenge the status quo of intergroup relations. In Israel, where there is consensus that relations can improve only when sources of division are addressed, the emphasis is on raising awareness and understanding of difference. Anxiety and threat are accepted as intrinsic dimensions of this process and mechanisms are developed to minimize their impact. By contrast, in Northern Ireland, where the achievement of school objectives is presented as a "by-product" of contact between Catholics and Protestants, there is seemingly less incentive to adopt a proactive approach to tackling difference. There is though a question around the extent to which it is really believed that, by a process of osmosis, the integrated schools will generate more tolerant and less prejudiced individuals. The evidence presented here suggests that some are unconvinced of this argument, but that it serves as a convenient rationale for not having to engage in intergroup activity that has the potential to induce conflict or tension that teachers feel ill-equipped to deal with.

Finally, the research presented here reinforces the point made by some social psychologists that contact in the real world does not happen in a vacuum. The nature of the intergroup encounter, social norms, and individual experiences all help to shape the contact experience and influence its effects. In Israel, for example, commitment to improving Arab/Jewish relations through the schools, alongside cultural norms of behavior that are accepting of openly expressed emotion and tackling problems of division "head on" facilitate a more conducive contact environment than is the case in Northern Ireland. In the Northern Ireland schools the seeming lack of commitment to focusing on division is compounded, if not causally related, to wider societal norms of conflict avoidance and polite, but essentially superficial, intergroup interchange. Factored into this, and at the institutional level, is the suggestion in our data that as integrated schools in Northern Ireland have proliferated they have lost some of their original sense of purpose. An exploration of why this has happened could form the basis of another paper, but there are some points worth highlighting here. First, unlike the bilingual, binational schools in Israel, integrated schools in Northern Ireland are evaluated using the same criteria that apply to segregated schools. Thus, whilst the Department of Education periodically inspects all schools in Northern Ireland to ensure adequate delivery of the national curriculum, there is no separate exercise to assess the operationalization and delivery of "integrated" objectives. Second, demographic change in recent years has informed the decision of many small (almost always Protestant) schools to transform to integrated status, where in the absence of increased pupil numbers, the schools are unlikely to meet viability criteria set by government. The fact that some schools are transforming as a survival strategy, raises questions about the commitment that many of them will have to the ethos of integrated education. Third, it was indicated to us that many teachers seek employment in integrated schools not because they espouse the ideals of integrated education, but because such schools, by virtue of being newly established, offer easier routes to promotion. Finally, unlike Israel, where dedicated teacher training and development is a key feature of bilingual/binational schools, there is little by way of comparison in Northern Ireland. Many teachers enter integrated schools having received little or no specialized training and the schools we examined did not see it as their responsibility to provide in-service training. Against the background of the ongoing, albeit faltering, peace process in Northern Ireland, there is perhaps much to be learned by policymakers responsible for the integrated agenda from the very dedicated approach to Arab/Jewish coeducation in Israel.

## Notes

The authors would like to thank the Nuffield Foundation for funding the research presented in this chapter and the referees who provided detailed and constructive comments on the first draft of it.

1. There are now 46 planned and transformed integrated schools in Northern Ireland accounting for around 5% of all schools at primary and secondary levels (Gallagher et al., 2003). In Israel there are just 4 bilingual/binational schools.
2. In reality few transforming schools are likely to attain this target in the short term as, by dint of previous segregated status, teachers and governors are likely to be substantially more representative of one community.
3. Partner teachers work together in every class to deliver lessons simultaneously in Arabic and Hebrew.
4. Rangers and Celtic are football teams that are strongly associated with the Protestant/Loyalist culture and the Catholic/Nationalist culture respectively. Sectarian rivalry is a feature of relations between these teams and their fans.

## References

Allport, G.W. (1954). *The nature of prejudice*. Reading, MA: Addison-Wesley.

Bar-Tal, D. (2000). From intractable conflict through conflict resolution to reconciliation: Psychological analysis. *Political Psychology* 21: 351–365.

Bekerman, Z. (2002). The discourse of nation and culture: Its impact on Palestinian-Jewish encounters in Israel. *International Journal of Intercultural Relations*, 26 (4): 409–427.

Blacking, J., Holy, L., and Stuchlik, M. (1977). *Situational determinants of recruitment in four Northern Irish communities*. London: Report to the Economic and Social Science Research Council.

Brown, R. and Hewstone, M. (2005). An integrative theory of intergroup contact. In M. Zanna (Ed.), *Advances in experimental social psychology*, vol. 37 (pp. 255–343). San Diego, CA: Academic Press.

Buckley, A. (1989). You only live in your body: Peace exchange and the siege mentality in Ulster. In S. Howell and R. Wallis (Eds.), *Societies at peace: Anthropological perspectives* (pp. 146–165). London: Routledge.

Department of Education, Northern Ireland. (2004). *Transformation, an information pack for schools*. Retrieved October 5, 2004 from http://www.deni.gov.uk/schools/pdfs/transformation_pack.pdf

Donnan, H. and McFarlane, G. (1983). Informal social organisation. In J. Darby (Ed.), *Northern Ireland: The background to the conflict* (pp. 110–136). Belfast: Appletree.

Geertz, C. (1973). *The interpretation of culture*. New York: Basic books Inc.

Hewstone, M. (1996). Contact and categorization: Social psychological interventions to change intergroup relations. In N. Macrae, C. Stangor, and M. Hewstone (Eds.), *Stereotypes and stereotyping* (pp. 323–368). New York: Guildford.

Hewstone, M., Rubin, M., and Willis, H. (2002). Intergroup bias. *Annual Review of Psychology*, 53: 575–604.

Hughes, J. (1992). The formation and management of national identity and intergroup attitudes amongst children in two polarized Belfast communities. Belfast: Queens University, Unpublished PhD thesis.

———. (2003). Resolving community relations problems in Northern Ireland: An intracommunity approach. *Research in Social Movements, Conflict and Change*, 24: 257–282.
Larsen, S. (1982). "The two sides of the house": Identity and social organization in Kiloborney, Northern Ireland. In A. Cohen (Ed.), *Belonging* (pp. 132–156). Manchester: Manchester University Press.
Leyton, E. (1974). Opposition and integration in Ulster. *Man*, 9: 185–198.
Northern Ireland Council for Integrated Education (NICIE). (2004). *Aims and Objectives*. Retrieved October 5, 2004 from http://www.nicie.org/aboutus/
Pettigrew, T.F. (1997). Generalized intergroup contact effects on prejudice. *Personality and Social Psychology Bulletin*, 23(2): 173–185.
Pettigrew, T.F. and Tropp, L. (2000). Does intergroup contact reduce prejudice? Recent meta-analytic findings. In S. Oskamp (Ed.), *Reducing prejudice and discrimination: Social psychological perspectives* (pp. 93–114). Mahwah, NJ: Erlbaum.
———. (2006). A meta-analytic test of intergroup contact theory. *Journal of Personality and Social Psychology*, 90(5): 751–783.
Tausch, N., Kenworthy, J.B., and Hewstone, M. (2005). The role of intergroup contact in prejudice reduction and the improvement of intergroup relations. In M. Fitzduff and C.E. Stout (Eds.), *Psychological approaches to dealing with conflict and war*, vol. 2: *Group and social factors* (pp. 67–108). Westport, CT: Praeger.

# Section III
# Curriculum and Pedagogy

# Chapter Ten
# Education for Peace: The Pedagogy of Civilization

*H.B. Danesh*

## Introduction

Peace and education are inseparable aspects of civilization. No civilization is truly progressive without education and no education system is truly civilizing unless it is based on the universal principles of peace. In reviewing school textbooks and theories upon which their contents are based, we find that the textbooks are predominantly written from the perspective that conflict and violence are inevitable and necessary aspects of individual and social life. They either inadvertently or deliberately promote a culture of violence and war (Frier, 2002). Consequently, new generations are taught by their parents, teachers, and community leaders the ways of "otherness," conflict, and violence. Seldom do we encounter a systematic educational program that teaches children and youth the principles of peace. This chapter briefly puts forward the Integrative Theory of Peace (ITP) and outlines the Education for Peace (EFP) Program developed on the basis of that theory.

The main premise of ITP and the EFP program is that all human beings relate to themselves, the world, and life through the lens of their specific worldview. It further holds that education has a cardinal role in the formulation and development of our respective worldviews in the context of family, school, and community. The main theme of the EFP program is that effective and sustained peace education needs to focus on all aspects of human life: intellectual, emotional, social, political, moral, and spiritual. Therefore, EFP introduces certain concepts and perspectives that go beyond the usually held views on peace education. (See for example, Salomon and Nevo, 2002).

## The Integrative Theory of Peace

The Integrative Theory of Peace consists of four subtheories:

- *Subtheory 1*: Peace is a psychosocial and political as well as a moral and spiritual condition;

- *Subtheory 2*: Peace is the main expression of a unity-based worldview;
- *Subtheory 3*: The unity-based worldview is a prerequisite for creating both a culture of peace and a culture of healing;
- *Subtheory 4*: Comprehensive, integrated, and lifelong education within the framework of peace is the most effective approach for a transformation from the metacategories of survival-based and identity-based worldviews to the metacategory of unity-based worldview.[1]

Based on this theoretical formulation, the EFP curriculum is designed to be comprehensive, integrative, all-inclusive, and simultaneously both universal and specific. The EFP curriculum is "comprehensive" and "integrative" in the sense that it considers all aspects of peace—biological, psychological, social, historical, ethical, and spiritual—and integrates them into one coherent and all-inclusive framework for the study of all school subjects. The "all-inclusive" aspect of the curriculum refers to the fact that all members of the school community—students, teachers, administrators, support staff, and indirectly, all parents—are included. The "universal" principles of peace, as formulated in the EFP curriculum, are fourfold: humanity is one; the oneness of humanity is expressed in the context of diversity; unity in diversity is a prerequisite for peace; and peace requires the ability to prevent and resolve conflicts without resorting to violence. While these principles transcend cultural/ethnic differences, their application within every community is "specific," as it aims to safeguard and celebrate unique cultural heritage within the context of these "universal" principles.

## Education for Peace Experiment in Bosnia and Herzegovina

In September 1999, the Federation of Bosnia and Herzegovina Minister of Education invited me to bring the EFP program to schools in Bosnia and Herzegovina (BiH) as an integral part of the European Community's contribution to the 1995 Dayton Peace Accord.[2] In May 2000, after consultations with BiH educational officials, school directors and teachers, as well as the Office of the High Representative (OHR)—the International Community Representative in BiH—the project was launched with a grant from the Government of Luxembourg covering the first year of the two-year pilot project. The second year was implemented primarily through voluntary work of EFP faculty, staff, and the participating BiH schools. This pilot project was administered in six schools (three primary and three secondary) representing the three main ethnic groups of BIH: Bosniak (mainly Muslim), Croat (mainly Catholic), and Serb (mainly Orthodox Christian) and all segments of the community—rich and poor, residents and the displaced, war victims and war participants, and urban and rural communities. The project involved approximately 6,000 students, 400 school staff (teachers, administrators, support staff), and 10,000 parents/guardians.[3] Participation was voluntary.

*Context: Bosnia and Herzegovina 1980–95*

In 1980, following the death of Marshal Tito, Yugoslavia slid into economic and political decline. The conditions of interethnic collaboration and harmony, imposed through Tito's authoritarian brand of Marxist ideology, were soon replaced by conflict and power struggle among Yugoslavia's ethnically diverse leaders. Eventually, a destructive doctrine of Serb nationalism promoted by Slobodan Milosevic became the dominant force in the region, resulting in a long and horrific war.

In Sarajevo alone, in the course of three long winters, more than 12,000 residents including 1,500 children were killed.[4] The barbarous July 1995 "ethnic cleansing" massacre of more than 6,000 men and boys of Srebrenica left a lasting blot on the moral conscience of the International Community (PBS, 1999). It is not possible to convey here the full extent and depth of the war tragedy in BiH. The following is one example of many devastating accounts:

> [I]n a country of 4.4 million inhabitants, it is estimated 250,000 (mostly Bosniaks) were killed and more than 240,000 injured, including 50,000 children. More than 800,000 were externally displaced and are now living abroad as refugees, and more than one million people were internally displaced and are now living in BiH in places other than their home communities. Between 10,000 and 60,000 women, primarily Bosniaks, were subjected to rape and other atrocities difficult to understand and to recount.[5]

The economic damage to BiH is staggering, and the recovery costs are estimated at USD10 billion (World Bank, 2000). The war ended with no peace in sight and with conflict still present. The solution was far from satisfactory.

Since the signing of the 1995 Dayton Peace Accord that ended a four-year of all-out war, a process of social, economic, and political recovery mainly initiated and administered by the International Community through financial and technical assistance and reconstruction projects has been taking place in BiH. However, relatively little attention has been given to education reform, and only recently have changes begun to be implemented (Ministry of Education, 2001). Ethnic and religious tension in BiH remains widespread and extreme nationalism has not yet been excised from the body politic. Broadly speaking, a culture of peace has not yet taken hold.

The groups particularly distressed by the war in BiH and its aftermath are children and youth of all ethnic backgrounds. Having experienced atrocities, violence, pain, and extremely difficult living conditions, many children and youth as well as their parents have been severely traumatized. According to one survey, an alarming 62% of Bosnian young people would leave the country if they had the chance (UNDP, 2000, p.11).

The educational system of BiH currently suffers from a number of weaknesses such as outdated teaching methodologies, irrelevant teaching materials, rote learning, rigid and overloaded curricula, and a pedagogical approach that does not "promote the development of students' potential."[6]

## EFP Program Description

*Conceptual Considerations*
ITP identifies four conditions for a successful peace education program: a unity-based worldview,[7] a culture of peace, a culture of healing, and a peace-based curriculum for all educational activities. Based on these conditions, EFP has four main goals: (1) to assist all members of the school community to reflect on their own worldviews and to

gradually try to develop a peace-based worldview; (2) to help participants to create a culture of peace in and between their school communities; (3) to create a culture of healing designed to gradually, but effectively, help members to recover from the damages of war and protracted conflict affecting their families, community members and themselves; and (4) to learn how to prevent new conflicts and resolve them in a peaceful manner, without resorting to violence, once they occur.

*Worldview, Education, and Peace*

The EFP program postulates that all conscious deliberate human activities, including the creation of conditions of conflict or peace, are shaped and determined, to a considerable degree, by our *worldview*[8] that itself is the outcome of the education received from our families, schools, communities, and accumulated life experiences. Thus, a comprehensive program of peace education requires attention to the welfare of the family, modes of parenting, school curriculum, pedagogical methodology, community relationships, economic conditions, sociopolitical policies, and leadership practices. In essence, true education is a humanizing process with the ultimate aim of creating a civilization of peace. In the EFP curriculum, three distinctive, but interrelated and developmental worldviews are identified: survival-based, identity-based, and unity-based.

The survival-based worldview uses power for domination and control and often exhibits authoritarian human relationships and modes of governance. This worldview characterizes earlier stages of development in both individual and collective domains of life and is prevalent in times of crises and danger, such as natural disasters, terrorism and war. Survival-based worldview has existed since the dawn of history and remains the dominant worldview in the context of many families, institutions, communities, and governments.

The identity–based worldview illustrates the coming of age of individuals, institutions, and societies alike. The primary foci of identity-based worldview are survival on one hand, and competition and winning, on the other. In this mode, participants engage in constant power-struggle and competition for individual and group advantage in all departments of life—personal, familial, social, economic, political, and even scientific and religious pursuits. In this worldview, issues of individualism, nationalism, racism, and other concepts that separate people and groups from each other dominate. The welfare of oneself and one's group are usually given preferential status. Within the framework of survival- and identity-based worldviews, competition, conflict, and even violence are generally the norm rather than the exception.

From a psychological perspective, the formulation and adoption of a worldview represent the development of human consciousness at both individual and collective levels. As such, survival- and identity-based worldviews are both natural and expected mind-sets of childhood and adolescence. Following this assumption, and based on the dynamics of human psychological development (see for example, Erikson, 1968), we can conclude that the next step in the expansion of human collective consciousness will be the development of a new worldview that, at once, takes into consideration at least three fundamental peace-related issues: ensuring safety and security for all; encouraging individual and group achievement and distinction; and providing opportunities for a purposeful life in a unified environment.[9] This worldview is designated the unity-based worldview. Elsewhere, I have described it in the following manner:

> Within the parameters of [the unity-based] worldview, society operates according to the principle of unity in diversity and holds as its ultimate objective the creation of a civilization of peace—equal, just, liberal, moral, diverse, and united. The unity-based worldview entails the equal participation of women and men in the administration of human society. It rejects all forms of prejudice and segregation. . . . It ensures that the basic human needs and rights—survival and security; justice, equality, and freedom; and the opportunity for a meaningful, creative life—are met within the framework of the rule of law and moral/ethical principles.[10]

The EFP curriculum introduces this formulation of worldview (survival, identity, unity) to participants (teachers, students, parents alike) and invites them to reflect and discuss how they apply to the their lives and histories. This process is conducted with extreme care, making certain that participants focus only on their own worldviews. The objective is to help them to embark on a journey of *self-knowledge*.

*Culture of Peace and Culture of Healing*
The EFP curriculum, formulated within the parameters of a unity-based worldview, aims to help teachers, students, and staff to create a *culture of peace* (UNESCO, 1998) in their school community. A culture of peace requires full and sustained involvement of all community members; no one can be excluded. Among the main characteristics of a culture of peace is its ability to create an atmosphere of mutual trust, respect, and recognition. These conditions, in turn, decrease fears (Harris and Callender, 1995), open lines of communication (Tal-Or, Boninger, and Gleicher, 2002), facilitate intergroup dialogue (Littlejohn and Dominici, 2001; Ozacky-Lazar, 2002), and encourage the undertaking of joint activities, all of which are essential for assuaging the wounds of violence and the creation of a *culture of healing*. The all-encompassing character of unity-based worldview and cultures of peace and healing require that the school curriculum in its entirety be implemented within the framework of principles and mind-set of peace. Furthermore, the EFP curriculum requires that all attempts to understand, prevent, and resolve disagreements and conflicts in school be enacted within this framework. In other words, *peace must become the primary approach to life, rather than life becoming a never-ending search for peace*. If we are to create peace, we must become peaceful and peace-creating. Without these elements, peace is certain to elude us.

## The Main Elements of the EFP Curriculum

*Integrative and Inclusive*
The EFP program's approach to peace education is integrative. It helps participants to develop the necessary knowledge, capacity, courage, and skills to create violence-free and harmonious environments in their homes and schools. It offers special training for students as well as their teachers and parents. EFP focuses on the

education and empowerment of girls and women, encouraging equality with their male counterparts in all arenas of life. The program also pays particular attention to the training of boys and men, educating them on how to avoid abuse of power and aggression and violence—behaviors traditionally expected of men. The program includes a major leadership training component, with the aim of preparing the students—the future leaders of the society—to become peacemakers.

*Universal and Specific*

EFP is based on the "universal" principle that humanity has one common heritage expressed via a diversity of races, languages, and cultures and that our primary task is to create a civilization of peace. These principles are both scientific and moral and apply equally to all people, in all societies, and under all conditions. They reject the dichotomous and conflict-creating perspectives that consider some races or cultures superior to others.

The "specific" aspect of the EFP curriculum refers to the fact that it is specially designed for each new community with the full participation of its own educators and scholars. The curriculum is designed to encourage the development of identities that are both unique and global, so that participants will see themselves as agents of change and progress for their respective communities within the parameters of an increasingly global order.

In the course of its implementation, the EFP program ensures cultivation of local human resources, strengthens interethnic dialogue, and involves the participation of the entire school community. The program provides ongoing training and professional development for all school staff, enhances the creative dimension of the learning process, and through its peace events and other activities, reaches out to the community at large. The following anecdote reflects the discourse that takes place during the implementation of EFP in the schools:

> In Travnik, two schools—a secondary school with an overwhelming majority of Bosniak (Muslim) teachers and students and a primary school with an equally large majority of Croat (Catholic) population—while located in the same municipality had no formal or informal contact prior to the introduction of EFP program. In light of their collaborative training in the principles of Education for Peace, teachers, administrators, and staff began to study the dynamics of the three prevailing metacategories of worldview: survival-, identity-, and unity-based. They then started to share their insightful thoughts and ask fundamental questions:

- During the time of Tito, we were all united; we intermarried; and we rarely focused on our religious affiliations. Why then did we engage in a barbaric war once the reign of Tito came to an end?
- What are the fundamental differences between the "unity based on sameness" promoted by Tito and the "unity in diversity" being taught by EFP?
- How could the leaders of two religions (Christianity and Islam), based on the principles of peace, love, and brotherhood, have consented to and even encouraged the recent barbarous wars and atrocities against each other's followers?

- What kind of assistance and training can we offer to the parents of our students who have nationalistic sentiments and continue to convey to their children lessons of dislike, distrust, and hatred against the members of the other religious group?

Through these questions, teachers attempted to apply the "universal" principles of peace to the "specific" context of their history and experience.

*Peace as the Framework*
In the EFP program, students are encouraged to apply the principles of peace to each subject matter and to try and develop new insights into that subject within the parameters of a unity-based worldview. For example, when students begin to study geography, the fact that the earth is fundamentally one and environmentally indivisible becomes evident. Most history texts are accounts of conflicts and wars conducted by men, with the attendant triumphs and defeats. In these historical accounts, peace is nonexistent and women are absent. However, when the same history is viewed from the perspective of peace, it becomes evident that all progress, development, and community building in any given society takes place during periods of peace and that women, minorities, and the ordinary masses make unique and outstanding contributions to the overall progress of their communities. When students study biology, they begin to appreciate the power of organic unity in the context of diversity that makes both life and living possible. The following profile of one of the participating schools demonstrates both the necessity for and the challenge of teaching the students how to apply the universal principles of peace to all aspects of school life:

> The Third Elementary School in Ilidzia (adjacent to Sarajevo) has a student population of more than 1,200, the majority of whom are displaced persons, refugees, and orphans—all living in extreme poverty and deprivation. The teachers and staff are likewise victims of the atrocities of the recent war. When the EFP program was offered to the school, they voluntarily and enthusiastically joined the pilot project. But when they became aware of the universal principles of peace, they wondered:

- How could they teach such lofty and seemingly unrealistic principles to students, other teachers, and parents who have been the victims of atrocities at the hands of former friends, neighbors, and fellow citizens?
- How could they speak of the nobility of human nature in the face of such ignoble acts?
- How could they talk about justice to their students given the unbelievable injustices they have suffered?

On the occasion of the first national peace event that was to take place in Banja Luka—the city of their "former enemies"—the level of fear, anxiety, and even revulsion at this school was so high that it required extraordinary measures of self-control and resolve to take some of their students, teachers, and parents to the other city to participate in the peace event. At the conclusion of that one-day peace event, participation in several other such events, and two years of intensive application of EFP, this school has become one of the most outspoken and enthusiastic promoters of EFP. They have found basic and practical answers to some of their most perplexing questions.

*Creating Culture of Peace and Culture of Healing*
The initial intensive phase of the project in each school lasts two years. During the first year students are taught about the characteristics and dynamics of a culture of peace. The second year is devoted to the continuation of the tasks of the first year, as well as to learning about and participating in the process of creating a "culture of healing" within and among their respective school communities. The culture of healing is a process designed to help whole populations of individuals, both adults and children, victims as well as perpetrators, to overcome the aftereffects of severe psychosocial, as well as moral-spiritual trauma from intractable conflicts, violence, and war. This approach is distinct from posttraumatic psychological interventions that are individual-centered and usually do not engage the "other" in the recovery process. As a result, while they reduce the emotional anguish of the individuals involved, they may also increase the anger and hostility toward those who are viewed as the perpetrators of the "wrongs" suffered by the subject. These programs are extremely expensive, require a large number of well-trained local experts, and are very time-consuming (Herman, 1992).

Another approach used for the recovery of postconflict communities is institution of the Truth and Reconciliation Commission (TRC) (Bloomfield, Barnes, and Huyse, 2003) that has been at work in several countries with varying degrees of success. The TRC approach has been considered, but not implemented in BiH, and in our view would not be appropriate for this society. The main reason, aside from reservations identified in the Truth and Reconciliation Commission review literature, is the fact that the TRC is both culture- and ethnic-specific, with reliance on public confession and forgiveness. Such practices are unsuitable for cultural settings that do not universally sanction them as a part of their religious orientation.

The EFP concept of the culture of healing[11] is based on the definition of "societal health" as a state of unity in its fullest sense—intrapersonal, interpersonal, and intergroup. A unified society has inherent healing properties at both individual and collective levels. Its creation is possible only within a framework characterized by unconditional recognition and celebration of diversity, mutual acceptance, interpersonal and intergroup empathy, social justice, freedom from all forms of prejudice, and equality of opportunity. One event that depicts aspects of a culture of healing is recounted here:

> The day before an intensive EFP training weekend for the staff (teachers, administrators, and support staff) of the two participating schools in Travnik, we were informed that the Croat primary school's staff would be a few hours late because of the untimely death of the mother of three young students at their school. This was a particularly tragic event given that these children had already lost their father and the whole school community was attending the funeral. Once the Croat school staff arrived, instead of focusing on the previously determined subject, a talk was given on "untimely death" and then for two hours the eighty participants, engaged in discussion of untimely death and its social, emotional, and spiritual impact on all involved.

Midway through the session, one of the Bosniak (Muslim) schoolteachers turned to his Croat (Catholic) colleagues across the aisle and said, "During the war my sister was caught behind the enemy [Croat] lines of war and became extremely sick. A few Croats risked their lives and drove my sister several hundred miles to a hospital in Zagreb for treatment." Then with tears streaming from his eyes, he said, "I never said thank you for this act of extreme kindness."

The Croat staff with an equal level of sincerity responded by sharing their accounts. One Croat teacher expressed her gratitude for those Bosniak families who, with selfless courage in the midst of great danger, hid and cared for a family of Croats at a time when intense ethnic conflict raged in that area. This exchange, characterized by a profound level of sincerity, sensitivity, and mutual admiration, completely removed the strangeness and aloofness that had till then existed between the two groups. A foundation of a *culture of healing* was being laid within and between the two schools.

## Project Organization

The six schools that participated in the pilot project received the "intensive" version of the EFP program. This version requires two years of almost daily communication, consultation, and collaboration between the school and the EFP on-site faculty. In addition, other members of the EFP Faculty, including one senior faculty and several senior consultants from the fields of education, psychology, law, conflict resolution, and public relations were intimately involved with the program and made frequent visits to the schools.

The international component of EFP on-site faculty—six young university graduates from Burkina Faso, Canada, Turkey, and the United States—were placed as two-person teams in each of the three cities (Banja Luka, Sarajevo, and Travnik) Each team was responsible for two EFP schools. They worked with a group of 5–8 teachers who together made up the 24 person on-site faculty of the EFP-BiH project.

The senior faculty prepared, for the training of EFP on-site faculty, extensive materials, including reading packages on the history of the Balkans and the psychological, political and economic aftereffects of the 1992–1995 war. During the training, lectures were given by specialists in political science, history, psychology, human rights law, the International Tribunal, curriculum development, and events planning. Throughout the year, additional support was provided by the senior faculty and the consultants with particular regard to the development of curriculum resources, organizational structures and procedures, and the formal evaluation of project activities.

After receiving intensive 8-day training in July 2000, the on-site faculty helped train the 400 BiH school staff and teachers at two-day seminars held in September and March. In addition, they provided throughout the year, regular training opportunities for teachers and students and conducted workshops aimed at deepening the understanding and application of the EFP conceptual framework materials. The on-site faculty members were also present in the schools on a *daily* basis, consulting with individual teachers, observing classes, providing assistance to student groups in their preparations for presentations at the "Peace Events," and giving encouragement for other peace-related initiatives.

## Process and Results of the Pilot Phase

*Initial Training of On-Site Faculty*
The initial training of on-site faculty began with an introduction to the concept of worldview. The impact of worldview awareness on the participants became evident during the first two days of the training. As the concept was presented and the three dominant worldviews described,—survival-, identity-, and unity-based—the participants immediately responded. Some were defiant, agitated, and angry. Others responded with a sense of optimism that they had discovered a new framework through which they could make sense of their lives and experiences. The majority, however, was perplexed and struggled with a new insight that was at once both reassuring and challenging. We have observed similar response patterns throughout the five-year application of the EFP program that involves an additional 106 schools and thousands of students and teachers.

Those who were actively involved in the recent conflicts and wars experienced the most difficulty with these concepts. In particular, individuals who were aggressors during the war were greatly challenged. It took six months of study, dialogue, and program implementation as well as personal and group contemplation, before the majority of these individuals were able to begin reviewing their individual and collective worldviews. By contrast, the children and youth were much more receptive aside from some of the children of authoritarian parents, who displayed a strong proclivity toward conformity and blind obedience (Danesh, H.B., 1978a).

At the other end of the spectrum were victims of the recent war—individuals who had lost family members and friends, been personally attacked and violated, and suffered severe physical and financial damage. These individuals primarily responded in two ways: either with a sense of victimization and demand for "justice, punishment, apology, and reparation," or with new insights that allowed them to view others in a more understanding and even forgiving manner (McCullough Worthington et al., 1994, 1997, 1998; McCullough, 2000; Staub and Pearlman, 2001). Another major element of this initial training was expression of considerable skepticism on the part of the participants. These trends—worldview challenge and initial skepticism—were found in many teachers and school staff in subsequent trainings.

*Training of Teachers and Staff*
All school staff, directly or indirectly involved with the students, receive training in the principles of EFP during three intensive 2-day training sessions at the start, midway, and end of each academic year. In addition, about 10–15% of the teachers in each school receive two years of on-the-job extensive training as EFP specialists. These specialists spearhead the continuing implementation of the EFP Program after the completion of the initial two-year phase. The content and pedagogical principles of the EFP specialization training are included in the EFP Curriculum and are now near completion.

*Initial Skepticism*
Initial skepticism toward the project began with a period of questioning by those involved in the endeavor. They questioned the viability of peace and asked whether

human beings are by nature aggressive. They wondered how to rear their children, particularly boys, according to the principles of peace. They expressed guarded hope that women could become active agents of change in their communities as a result of participation in the EFP Project. They wondered how domestic abuse could be stopped and how the foundations of marriage and family could be strengthened. They also posed questions about the root causes of interethnic and interreligious animosities; the values and pitfalls of democracy; the relationship between economy and peace; the need for ethical governance; and many other significant issues that were relevant to their unique conditions. Most notably, students, teachers, and others expressed understandable skepticism about the validity, viability, and efficacy of the EFP Project as well as its objectives, given the historical and current realities in BiH.

*Gradual Acceptance*
The initial skepticism soon gave way to general acceptance of the program. Six months after the project's introduction, its positive and "transformative"[12] impact was so widespread that the project received an invitation from the Government of BiH to develop a plan to introduce the EFP Project in all elementary and secondary schools in the country. Likewise, the OHR called for the introduction of EFP in as many schools as possible in BiH and neighboring countries. In 2003, the EFP program was introduced in an additional 102 secondary schools in 65 communities across BiH.

EFP is one of the few peace education programs that has been voluntarily adopted by all three BiH ethnic communities and their respective educational authorities, without any change in its fundamental components. The EFP program cuts across differences that previously were obstacles to the creation of peaceful relationships. It has received enthusiastic support from the BiH Government and in May 2002 its Mission at the United Nations in New York, addressed an open letter to the Special Session of the United Nations General Assembly on Children (May 8–10, 2002), describing the pilot EFP program in the six BiH schools, stating that "the results of this program have been very positive," and concluding their letter with this call:

> The children all over the world are in need of peace and security. On the occasion of the Summit devoted to the children, we recommend this program [EFP] to all the nations for consideration, as a model of society oriented towards peace, cooperation, and development.

The Office of the High Representative, the ultimate authority of the International Community in BiH, has supported EFP on the basis of its design and impact:

> This is a unique project. It will teach how to create a violence-free environment, in homes and schools and in the country as a whole. (Ambassador Dr. Matei Hoffmann, The Senior Deputy High Representative, June 28, 2000)
>
> This invaluable project was conceived in such a way that the soul-searching process of reflection which the participants undergo as the project unfolds—be they pupils, teachers, parents, administrators, ordinary school workers—results, largely speaking, as we have ascertained ourselves, in a heightened holistic awareness of the war period and

its tragic consequences, and indeed triggers the desire amongst them to become authentic peace-makers, and precisely provides them with the necessary tools to achieve this goal . . . (Claude Kieffer, Senior Education Advisor, Office of the High Representative)

*Worldview and Attitude Change*
In the BiH pilot schools, EFP's main impact pertains to the attitude and behavior changes of children, youth, and adults participating in the program. The relationship between worldview and human conditions of conflict and peace was empirically demonstrated. The following excerpts provide a few examples of the nature and quality of this transformation:

> This project has changed our vision and worldview. I feel that the vision of every teacher and students in this school has been in some way changed through this project. (Bosnian Literature Teacher, 2nd Gymnasium, Sarajevo)
>
> As a result of participating in the EFP project, my way of teaching has changed, my relationships with students has changed, and my relationship with my family has changed . . . all for the better. (Teacher, Mixed Secondary School, Travnik)
>
> We should be able to recognize a good opportunity, which would help us to create justice and equality. Education for Peace is that chance and it gives us some very good instructions. This project helps us to become peaceful and optimistic. This project puts us in the position to use our own will and best thoughts to make human life better. (Student, Banja Luka Gymnasium)

*Pedagogical and Curriculum Reform*
The EFP program initially did not have a plan to reform the education curriculum of the participating schools. However, early in the project it became evident that the prevailing pedagogical approaches in BiH needed to be updated with focus on active student participation in their learning endeavors. It was also evident that in order for the EFP program to be effective, a more "democratic classroom learning" milieu was required. In response to these needs, a specially designed pedagogical approach was created with the assistance of two senior educators from Canada and the active participation of BiH teachers and pedagogues.

The main objective of this new approach was to equip teachers with the skills to create integrative lesson plans for the purpose of training students to recognize the principles of peace in and across all areas of study. All teachers were provided with a planning template for the preparation of "understanding-oriented" and "process-oriented" lessons. This approach enabled teachers to integrate EFP concepts into their lesson plans while inviting the active participation of students in the learning process. The following comments by the teachers and students provide a few examples of the impact of the program on the schools' pedagogical approaches and curriculum:

> Before this project, things were imposed in our classes, but with EFP we do it because we love it. (Student, Nova Bila Primary School)

The EFP project has helped us look at our syllabus in a different way, from a different perspective, giving us a chance to enrich it with issues not dealt with so thoroughly before. Although it hasn't always been easy, especially at the beginning, I think that we have become more confident in applying the principles of peace. (English Teacher, Mixed Secondary School, Travnik)

This is a good project because it gives students an opportunity to express themselves in a different way from what we have done and through creativity and the arts. They try to show us how a peaceful society can be. Through the presentations, they raised an understanding between students, teachers and parents. (2nd Gymnasium Math Teacher, Sarajevo)

*Interethnic Reconciliation*

If we were to single out the most important achievement of the project during its first two years, it is that an ethnically diverse group of students and teachers began a process of meaningful and sustained reconciliation and friendship. During this period the level of interethnic comfort increased dramatically. For the first time since the 1992–95 war, both adults and children traveled to the cities of their former combatants. Many started a process of regular contact and communication through e-mail, telephone, and personal visits; and a new sense of mutual trust and acceptance pervaded the whole EFP community. These assertions are based on a large number of individual interviews, questionnaires given to all students, reports prepared by teachers and administrators, and observations offered by the parents and other community members who were closely involved in the school community. It remains to be seen what the long-term effects of this transformation will be, but they have clearly evolved and intensified during the past five years, since the inception of the program. Here are a few statements by teachers, parents and students about the process of establishing new bonds of friendship across various groups:

The EFP project has brought some changes to our school, our community and our families . . . The collaboration between parents and the school has become better, and the teachers and their parents from Travnik [for the first time since the war] have visited our school. (Parent and Support Staff, Nova Bila Primary School)

I was pleasantly surprised with the way in which the pupils accepted this project. They accepted it very seriously and they have shown a great deal of interest and creativity through presentations that have shown their vision of peace and unity. My opinion is that we should spread this project and put more energy into it because the children are smart and they can do a lot for this world. (Geography Teacher, 3rd Primary School, Ilidza)

I think that this EFP is a very important thing for young people to understand the importance of peace. (Grade 11 student, Banja Luka Gymnasium)

I think that a main part of the EFP project is to share our understanding of peace and to learn how to become peacemakers. It really doesn't matter where you are from. I thought in the beginning that this project wouldn't affect anyone, but to me the effect has been so amazing. Students have been so excited to be involved in it. (Grade 9 student, Mixed Secondary School, Travnik)

Despite positive feedback, we do not consider the task of the project to be completed. We see an ongoing focus within three specific areas in the EFP schools as essential: (1) maintenance and acceleration of efforts to establish a well-grounded culture of healing; (2) further strengthening of the culture of peace already established between thousands of students, teachers, and parent; and (3) permanent inclusion of EFP in the curriculum of the participating schools.[13]

## Evaluation of the Results

The most intensive and far-reaching external evaluation of the program, thus far, has been conducted by a team of two experts commissioned by the Swiss Development and Cooperation Agency (SDC), which has given a major grant for the implementation of the EFP program in 100 secondary schools in BiH. This evaluation was performed by SDC halfway through the four-year duration of the grant and is dated September 2004. Here are the main observations and conclusions;

*Interaction of the Project with Ministries and Pedagogical Institutes*
It must be looked at as an achievement that all of the 13 Ministers of Education had agreed to participate in this EFP program as well as the Directors of the 8 pedagogical institutes and one hundred directors of secondary schools. The Ministers, Deputy Ministers, Directors of Pedagogical Institutes and Directors of secondary schools, met by the evaluation team, talked positively about the program, though few of them had attended any project seminars. For them the project led to a spectacular peace event and to children and teachers getting together.

*New Didactical Elements and Learning Experience*
School directors and school principals were seldom themselves confronted with the didactical material of EFP. They made reference to the work of the teacher when asked about the effectiveness and didactical learning methods used by the EFP program. The teachers interviewed first and foremost mentioned the opportunity to be trained by the EFP program through a new educational framework which offers didactical possibilities:

- more interaction between students and teachers,
- an open forum for discussion between students and teachers, and
- the relief of not having to load students with drill exercises.

To be a "peacemaker" was declared an important learning target by many teachers. Almost all students interviewed referred to the impact of EFP in positive terms. Some students also mentioned that EFP had been a common topic to discuss with their parents.

*The Importance of Bringing People Together*
The most important part of the project seems to be that it has brought people together across nationalities and languages; it has provided a place to meet. Several persons said

that in the education sector there was no other project like this. It provided and continues to provide physical spaces and opportunities for people to meet, share their experiences and build friendships. The project seems to have had—and continues to have—a healing effect on a war-torn nation. One of the teachers said: "The biggest impact was on the psychological level. People got an opportunity to express their emotions. We need this type of therapy. It had to do with the atmosphere created."

*Impact on Teachers, Staff and Students*
There is little doubt that the project has had great impact on many of the participants—teachers, support staff, administrators and students. The most important impact seems to have been on the personal level, the meeting of people across nationalities and languages.

*Conclusions by SDC [Swiss Development and Cooperation Agency]*
It may be concluded from the evaluators' observations and comments that the program is generally well received by Bosnian pupils, teachers and authorities. Psychological elements such as "bringing people together in an atmosphere of trust" as well as a number of didactical innovations are recognized by the evaluators. EFP has achieved positive impact not only among teachers and students but also among the families of participants.

However, while the value of EFP modules and the e-learning component—from a point of view of learning contents and didactics—has been well established, program effects on the behavior of students, teachers and communities at large can be assessed only at a later date and will present certain methodological challenges.

These observations are in harmony with the internal evaluations conducted by the EFP team. In order to identify the impact of the program and the apparent reasons for its success, we conducted random interviews, reviewed frequent reports provided by each segment of the school population, and made first hand observations of participants' responses to the program. The majority of participants and all of the senior education officials have pronounced the EFP program as highly successful.

## Critical Considerations

At the empirical level, the EFP program has demonstrated its effectiveness as a peace education program in a highly conflicted society, having recently emerged from a barbaric interethnic war. However, there are several issues and challenges that need consideration, particularly because the EFP Program is now drawing increasing attention from civic and governmental organizations in a number of countries. Among the most important of these challenges are conceptual considerations, implementation issues and research issues.

*Conceptual Considerations*
The primary challenge of the Education for Peace Program concerns conceptual formulations. If we consider peace to be the outcome of unity and conflict to be the

absence of unity, the EFP program calls for a totally new understanding of both the nature of conflict and peace. We consider conflict and violence symptoms of the underlying state of disunity. Therefore, conflict resolution and peace-making are both processes of unity-building. It is here that the unity-based worldview, which constitutes the framework of the EFP curriculum, assumes its primary significance.

In the EFP curriculum, peace is defined as a psychological, social, political, moral, ethical, and spiritual state. To create peace we need to consider all these issues and to educate new generations who not only understand the principles of peace, but also embody these principles in their personal lives, interpersonal relationships and social responsibilities and actions. Thus, the EFP approach to peace is at once psychological, social, and spiritual and as such it is at odds with many of the current theories and perspectives on peace that do not integrate the spiritual aspect of peace in their formulations.

*Sustainability*
An important challenge before the EFP program is how to secure long-term sustainability. To meet this challenge we plan to:

- Introduce the principles of education for peace into the curriculum used to train new teachers;
- Offer on-going training for current teachers and school administrators and staff;
- Gradually replace components of the school curriculum written within the framework of conflict-based worldviews with new text books and lesson plans prepared with due consideration of the principles of peace.

These objectives require significant changes to the currently held views on the nature and purpose of education, education policies, teacher-training curricula, allocation of financial resources, and political and ideological sensitivities surrounding the issue of education in every society. These issues are currently being addressed in Bosnia and Herzegovina in a systematic manner in close collaboration with education authorities.

*Financial Considerations*
Another issue to consider is the *cost* of the program, which is between USD100–200 per year per student for the "*EFP-intensive*" version of the program. At one level, this is a miniscule expenditure compared to the costs of war. On the other hand, it is considerable given the limited funds allocated to education particularly in economically disadvantaged societies. It should be mentioned that over 60% of this cost is for incentive payments to the local teachers and school staff who spend anywhere between 4–10 hours of additional work per week in order to implement the EFP program in their classrooms and schools.[14]

*Human Resources*
The ongoing expansion of the EFP program requires an increasing number of *EFP-trained educators and trainers*. As stated before, the EFP program has developed systematic training programs and strategies to deal with this issue.

*Time Considerations*
Another important issue is that *EFP is not a fast fix*. To alter a culture of conflict rooted in ancient and recent historical, religious, and ethnic hostilities requires effective, sustained transformation to a culture of peace within the parameters of a culture of healing. The curriculum needs to be changed and new generations of leaders and citizens need to be educated in the ways of peace. These are long-term and ongoing processes.

*Research and Evaluation*
Finally, it is evident that comprehensive research needs to be done on various aspects of the EFP program, the nature and dynamics of its impact on participants, and its long-term effects in educating new generations who are aware of the principles of peace, willing to implement them, and to become active agents for creating unity and peace.

Currently two longitudinal research projects on the EFP program in BiH are in progress, one conducted by a team of researchers from Columbia University (NY) and the other by EFP-International research team. This latter project is made possible through a grant from the United States Institute for Peace (USIP) and other sources.

*Modes of Delivery and Relevance of EFP*
*to Other Communities*
The success of the EFP program and its subsequent expansion has necessitated development of new pedagogical strategies in BiH. Concomitant with this process and in response to interest from educators and policy makers from other countries, we have developed variations of the program applicable to other groups and communities. Some examples include an EFP program for post-conflict and violence-afflicted communities and an EFP program entitled Violence-Free Schools designed specifically for schools located in more prosperous societies which are plagued by conflict and violence. We have also designed a comprehensive Web-based version of the EFP program intended for both primary and secondary school teachers and students and composed of a number of lessons in the fundamental concepts and elements of peace.

The attached appendix provides a list of the most import elements of the EFP program.

## Summary

Education for Peace (EFP) is a comprehensive and integrative program of peace education. Its primary focus is on the education of children and youth and involves teachers, students, parents, and the wider community. The main elements of the EFP program reflect the all-inclusive and integrative nature of peace itself. The EFP curriculum is universal in principle and specific in application. This objective is achieved through the active participation of educators and experts from every community in which the program is implemented. To ensure the sustainability of the program, during the first two years of its implementation, the project trains the necessary number of teachers in each school as EFP expert consultants. In the course of its implementation, EFP ensures cultivation of local human resources, strengthening of interethnic dialogue and collaboration, participation of the entire school community, on-going training and professional development of all school staff, study of the relevance of peace principles to all subjects, introduction of creativity in the learning process, and extension of the EFP program to the community at large.

Inherent in the concept of a culture of peace is the notion that peace is an expression and outcome of the integrated and comprehensive education of every new generation within the parameters based on the principles of a unity-based worldview. As such, education for peace constitutes the pedagogy of civilization in its truest sense and acts as the main instrument for training children and youth as peacemakers. Expressed differently, peace, education and civilization are inseparable dimensions of human progress.

The pilot EFP Project implemented in six Bosnia-Herzegovina schools has had considerable positive impact and is now being implemented in 106 new BiH secondary schools. The EFP Curriculum is being prepared in both print and multimedia formats. Both its distinctive conceptual framework and its large-scale application merit attention of educators and researchers interested and engaged in the subjects of peace and education.

## Appendix

A Check-List of Concepts and Activities of the Education for Peace Program

*A. Theoretical/Conceptual Issues*

I The Integrative Theory of Peace Education

- Subtheory 1: Peace is a psychosocial and political as well as a moral and spiritual condition;
- Subtheory 2: Peace is the expression of a unity-based worldview.
- Subtheory 3: Education is the most effective approach for development of a unity-based worldview;
- Subtheory 4: The unity-based worldview is the prerequisite for creating both a culture of peace and a culture of healing;
- Subtheory 5: Only a peace founded on a unity-based worldview is capable of meeting the fundamental human needs and human rights.

II Concept and Categories of Worldview

- Survival-based worldview
- Identity-based worldview
- Unity-based worldview

III Principles of Peace

- Humanity is one;
- Oneness of humanity is expressed in diversity;

- The primary challenge before humanity is to safeguard its oneness and celebrate its diversity; and
- To meet this challenge in a peaceful manner and without resort to violence.

IV The Concept of Human Needs and Human Rights

- Survival Needs (security, shelter, food, education, and so on) and the right to their fulfillment;
- Association Needs (equality, justice, freedom, and so on) and the right to their fulfillment;
- Transcendent Needs (meaning, purpose, righteousness, freedom of conscience, and so on ) and the right to their fulfillment.

V Special Issues

- The dilemma of power
- The question of authority
- The concept of unity
- Conflict-free conflict resolution

*B. Education for Peace Curriculum*

I Prerequisites

- Elements of a unity-based worldview
- Elements of a culture of peace
- Elements of a culture of healing
- Peace as the framework for the curriculum

II Components

- Study of the unity-based worldview
- Study of the elements of a culture of peace
- Study of the elements of a culture of healing
- Study of all subjects within the framework of peace

III Application

- Application of the unity-based worldview
- Creation of a culture of peace
- Creation for a culture of healing
- Creation of a peace-based curriculum

IV Characteristics of the EFP-Curriculum

- Comprehensive
- Integrative

- All-inclusive
- Universal
- Specific

V Pedagogical Considerations

- Training of all teachers, administrators, and support staff in the principles of EFP;
- Intensive Training of 10–15% of teachers/staff in each school as EFP specialists;
- Preparation of lesson plans by the teachers for every subject (biology, history, sports, math, . . .) according to the principles of peace and unity-based worldview;
- Holding schools-wide open-house, peace week at each school, every semester, involving the parents and the larger community;
- Holding regional peace events, once every semester, involving all EFP schools in the region;
- Holding national peace events, once a year, involving EFP schools representing all segments of the society;
- Creation of Youth Peacebuilders Network (YPN) Clubs in every EFP school, also involving youth from all the other schools;
- Use of multimedia production of the EFP curriculum on-line and CD-Rom;
- Facilitating live discourse and communication between EFP teachers, students, and parents wherever the required technical facilities are available.

## Notes

This chapter is originally a paper in series of papers based on a longitudinal research project on the EFP Program partially sponsored by a grant from the United States Institute for Peace (USIP Grant: SG-04003S). The author is indebted to Sara Clarke-Habibi, Associate Director, EFP-International, and Naghmeh Sobhani, Director of EFP-Balkans, for their outstanding contributions to the EFP Program. Special thanks are also due to Kimberly Syphrett and Gabriel Power for their excellent research work and to Christine Zerbinis and Stacey Makortoff for their invaluable editorial assistance.

1. H.B. Danesh, Towards an Integrative Theory of Peace, *Journal of Peace Education*, vol. 3, no. 1(March 2006): 55–78
2. http://www.cia.gov/cia/publications/factbook/geos/bk.html#People
3. In 2005, at the time of writing this chapter, the EFP program was being administered in 106 schools in BiH with a major grant from the Swiss Development and Cooperation Agency (SDC), and additional grants from Japan International Cooperation Agency (JICA), the Canadian International Development Agency (CIDA), the United States Institute for Peace (USIP), Rotary International, and Vectis Systems Inc (Canada). EFP also enjoys the full support of OSCE Mission in BiH and all 13 Ministries of Education of Bosnia and Herzegovina, who have called for the introduction of EFP to all BiH primary and secondary schools. According to the 2005 statistics provided by the office of OSCE in BiH, currently there are 370,114 primary school and 157,733 secondary school students registered in BiH.
4. Human Rights Watch (2002). *Human Rights Watch World Report 2002* www.hrw.org/wr2k2/pdf/bosnia.pdf (accessed October 2006).
5. Constance A. Morella, Congresswoman Morella War Crimes in Bosnia March 25, 1998 Congressional Human Rights Caucus Members' Briefing http://www.house.gov/lantos/caucus/index/archive/briefings/statements/testimony/cmorella3-25-98.htm (Morella, 1998; Vranic, 1996; UN Security Council, 1993; UCSF MEDICA Report, 1996).

6. May 2000 report of the Human Rights/Rule of Law Department of the Office of the High Representative.
7. H.B. Danesh (2002) Breaking the Cycle of Violence: Education for Peace. Included in African Civil Society Organization and Development: Re-Evaluation for the 21st Century. United Nations, New York, 32–39.
8. The foundation of every culture is its worldview, a concept that Moscovici (1993, pp. 160–170) calls "social representations," and Hägglund describes as "cultural fabric," stating that worldviews "constitute discursive complexes of norms, values, beliefs, and knowledge, adhered to various phenomena in human beings' lives" (Hägglund, 1999, pp. 190–207). Worldviews are usually expressed at a subconscious level (Zanna and Rempel, 1988; Van Slyck, Stern, and Elbedour, 1999; Guerra et al., 1997), and there is ample evidence that most peoples of the world live with conflict-oriented worldviews, be they ethnic-, religion-, or environment-based (Van Slyck, Stern, and Elbedour, 1999). Worldviews are also at the core of some of the current peace-related concepts and approaches such as story telling, (Bar-Tal, 2000) "contact theory," (Allport, 1954; Pettigrew, 1998), collective narrative (Rouhana and Bar-Tal, 1998, p. 762) and dialogue (Sonnenschein, Halabi, and Friedman, 1998; Suleiman, 1997).
9. It is the premise of the Integrative Theory of Peace (ITP) that basic human rights should reflect fundamental human needs. ITP posits that human needs are developmental in their process and biological, psychosocial, as well as spiritual in their nature. Within this framework, three basic categories of needs are identified: survival, association, and transcendent (supraordinate) needs. Of these needs, survival is the most immediate, association the most compelling, and transcendence the most consequential. Survival needs include issues of safety, security, and basic financial, health, shelter, food, and education requirements. Association needs refer to issues of human relationships such as equality, freedom, and happiness. Supraordinate needs are about issues of purpose, meaning, and spiritual convictions. For a detailed review, see H.B. Danesh, Towards an Integrative Theory of Peace . *Journal of Peace Education*, vol. 3, no. 1 (2006A): 55–78.
10. Danesh, Towards an Integrative Theory of Peace.
11. H.B. Danesh, Creating a Culture of Healing in Schools and Communities: An Integrative Approach to Prevention and Amelioration of Violence-Induced Conditions, forthcoming.
12. As expressed by the senior education consultant to the OHR in BiH.
13. All 13 BiH Ministries of Education and all 8 BiH Pedagogical Institutes have already endorsed the EFP program, and the plans for the training of all staff of Pedagogical Institutes in EFP are now under active consideration. In turn, these institutes will systematically incorporate the EFP principles in their teacher-training programs. These steps are essential if the effects of the EFP program are to be sustained and increased continuously.
14. Tim Houghton and Vaughn John of University of KwaZulu-Natal, South Africa in their thoughtful review of this article raise the following important questions: "Given the perceived failure of (massive) economic regeneration per se to effect reconciliation and harmony in Northern Ireland, could it be an economic regeneration approach has a better chance of contributing to peace if it is linked to payment and incentives for peace workers rather than relying on a stimulation of the broader economy? Such consideration has important implications for developing economies."

## References

Baruch, N. and Brem, I. (2002). Peace education programs and the evaluation of their effectiveness. In Salomon, G. and Nevo, B. (Eds.), *Peace education: The concepts, principles, and practices around the world*. Mahwah, NJ and London: Lawrence Erlbaum Associates.

Bloomfield, D., Barnes, T., and Huyse, L. (2003). *Reconciliation after violent conflict: A handbook*. In International Institute for Democracy and Electoral Assistance. Retrieved from http://www.idea.int/conflict/reconciliation/reconciliation_full.pdf

Danesh, H.B. (1969). "Growth" and psychotherapy. *The Chicago Medical School Quarterly*, 28 (3) 12: 75–86.

———. (1977a). Anger and fear. *American Journal of Psychiatry*, 134: 1109–12.

———. (1977b). The Angry Group, *International Journal of Group Psychotherapy*, 27(1): 59–65.

———. (1978a). The authoritarian family and its adolescents. *Canadian Psychiatric Association Journal*, 23(7): 479–484.

———. (1978b). In search of a violence-free community. *Mental Health and Society*, 5: 63–71.

———. (1979). The violence-free society: A gift for our children, *Bahá'í Studies*, 2, October.

———. (1980). The angry group for couples: A model for short-term group psychotherapy. *The Psychiatric Journal of the University of Ottawa*, 2: 118–124.

———. (1986). *Unity: The creative foundation of peace*. Ottawa and Toronto: Bahá'í Studies Publications and Fitzhenry & Whiteside

———. (1995). *The violence-free family: Building block of a peaceful civilization*. Ottawa: Bahá'í Studies Publications.

———. (1997). *The psychology of spirituality: From divided self to integrated self*. Switzerland: Landegg Academy.

———. (2001). Fever in the world of the mind: The nature and dynamics of violence. Monograph included in the EFP curriculum used in BiH schools.

Danesh H.B. and Danesh, R.P. (2002a). Has conflict resolution grown up?: Toward a new model of decision making and conflict resolution. *International Journal of Peace Studies*, 7(1): 59–76.

———. (2002b). A consultative conflict resolution model: Beyond alternative dispute-resolution. *International Journal of Peace Studies*, 7(2): 17–33.

Danesh, H.B. (2002c) Breaking the cycle of violence: Education for peace. In *African civil society organization and development: Re-evaluation for the 21st Century*, pp. 32–39. New York: United Nations.

Danesh, R. and Danesh, H. (2004). Conflict-free conflict resolution process and method. *Peace and Conflict Studies*, 11 (2): 55–84.

Danesh, H.B. (2004). *Peace moves*. Switzerland: EFP-International Publications.

Dayton Peace Accord (1995). *General Framework Agreement for Peace in Bosnia and Herzegovina*. Retrieved from http://www.mondediplomatique.fr/cahier/kosovo/dayton-en

Dayton Agreement (n.d.). Retrieved from http://www.cia.gov/cia/publications/factbook/geos/bk.html#People

Department for International Development Resource Centre for Health Sector Development (1999). *Bosnia Health Briefing Paper*. Retrieved from http:// www.healthsystemsrc.org/HBD/Text/Bosnia.doc

DePaul University (1993). War among the Yugoslavs. *Anthropology of East Europe Review*, 11(1–2). Retrieved from http://condor.depaul.edu/~rrotenbe/aeer/

Don Hays (2002, May). *Presentation by PDHR on behalf of the High Representative at a roundtable meeting of principals, parents and pupils from three model schools in the Sarajevo area*. Retrieved from http://www.ohr.int/ohr-dept/presso/presssp/default.asp?content_id=8335

Erikson, E. (1968). *Identity, youth, and crisis*. New York: W. W. Norton.

European Commission (2001). *Shared strategy for modernization of primary and general secondary education in BiH*, European Commission—Technical Assistance for Education Reform.

Frier, Ruth (2002). The Gordian knot between peace education and war education. In Salomon, G. and Nevo, B. (Eds.), *Peace education: The concepts, principles, and practices around the world*. Mahwah, NJ and London: Lawrence Erlbaum Associates.

Gavious, A. and Mizrahi, S. (1999). Two-level collective action and group identify. *Journal of Theoretical Politics*, 11(4): 497–517.

Haslam S.A, Powell C, and Turner J.C (2000). Social identity, self-categorization, and work motivation: Rethinking the contribution of the group to positive and sustainable organisational outcomes. *Applied Psychology: An International Review*, 49 (3): 319–339.

Harris I., and Callender, A. (1995). Comparative study of peace education approaches and their effectiveness. *NAMTA Journal*, 20(2): 133–145.

Herman, J. (1992). *Trauma and recovery*. New York: Harper & Row Human Rights Watch (2002). *Human rights watch world report 2002*. Retrieved from www.hrw.org/wr2k2/pdf/bosnia.pdf

Israeli, R. (1999). Education, identity, state building and the peace process: Educating Palestinian children in the post-Oslo era. In Raviv, A., Oppenheimer, L. and Bar-Tal, D. (Eds.), *How children understand war and peace*. San Francisco: Jossey-Bass Publishers.

Littlejohn, S.W., and Dominici, K. (2001). *Engaging communication in conflict: Systemic Practice*. Thousand Oaks, London, New Delhi: Sage Publications, Inc. May 2000 Report of the Human Rights/Rule of Law Department of the Office of the High Representative. Retrieved from http://www.ohr.int/ohr-dept/hr-rol/thedept/education/default.asp?content_id=3520

McCullough, M.E., and Worthington, E.L.J. (1994). Models of interpersonal forgiveness and their application to counseling: Review and critique. *Counseling and Values*, 39(1): 2–14.

McCullough, M.E., Worthington, E.L.J., and Rachal, K.C. (1997). Interpersonal forgiving in close relationships. *Journal of Personality and Social Psychology*, 73(2): 321–336.

McCullough, M.E., (2000), Forgiveness as human strength: Theory, measurement and links to well being. *Journal of Personality and Social Psychology*, 75 (6): 1586–1603.

McCullough, M.E., Rachal, K.C., Sanadage, S.J., Worthington, E.L.J., Brown, S.W., and Hight, T.L. (1998). Interpersonal forgiving in close relationships: II. Theoretical elaboration and measurement. *Journal of Personality and Social Psychology*, 19(1): 43–55.

Ministry of Education FBiH (Sept. 2001). Basis for education policy and strategic development in education.

Morella, Constance A. (1998). Congresswoman Morella War Crimes in Bosnia March 25, 1998 Congressional Human Rights Caucus Members's Briefing. Retrieved from http://www.house.gov/lantos/caucus/index/archive/briefings/statements/testimony/cmorella3-25-98.htm

Office of the High Representative (2000a). Agreement of the Conference of Ministers of Education of Bosnia and Herzegovina. May, 10, 2000.

Ozacky-Lazar, S. (2002). *Israel: An integrative peace education in an NGO—The case of Jewish-Arab centre for peace at Givat Haviva*. In Salomon, G. and Nevo, B. (Eds.), *Peace education: The concepts, principles, and practices around the world*. Mahwah, NJ and London: Lawrence Erlbaum Associates.

PBS (1999). *Srebrenica: A Cry from the Grave*. Retrieved from http:// www.pbs.org/wnet/cryfromthegrave/

Salomon, G., and Nevo, B. (Eds.) (2002), *Peace education: The concepts, principles, and practices around the World*. Mahwah, NJ and London: Lawrence Erlbaum Associates.

Staub, E., and Pearlman, L.A. (2001). Healing, reconciliation and forgiving after genocide and other collective violence. Helmick, S.J. and Petersen, R. (Eds.), *Forgiveness and reconciliation*. Randor, PA: Templeton Foundation.

Tal-Or, N., Boninger, D. and Gleicher, F. (2002). *Understanding the conditions and processes necessary for intergroup contact to reduce prejudice*. In Salomon, G. and Nevo, B. (Eds.), *Peace education: The concepts, principles, and practices around the world*. Mahwah, NJ and London: Lawrence Erlbaum Associates.

UNESCO (1998). Retrieved from http://www.unesco.org/cpp/uk/projects/infoe.html

United Nations Development Programme (UNDP) (2000). Independent Bureau for Humanitarian Issues and Prism Research, *Human Development Report—Bosnia and Herzegovina 2000: Youth*

World Bank (2000). World Bank Announces Strategy for Bosnia and Herzegovina. Retrieved from http://www.worldbank.org.ba/eca/bosnia&herzegovina.nsf/ECADocByLink/A3E92F806A42A41485256BB80077143D?Opendocument#1%20As%20a%20first%20step%20in%20this%20process

# Chapter Eleven
# Learning to Do Integrated Education: "Visible" and "Invisible" Pedagogy in Northern Ireland's Integrated Schools

*Chris Moffat*

Northern Ireland's integrated schools have attracted wide interest as one model of peace or multicultural education (Dunn and Morgan, 1999; UN CESCR, 2002). There is enough evidence to show that they "work" at least as well as segregated schools. They seem to address a presumed social and psychological reality: that social distance and lack of contact between Catholics and Protestants foster prejudice and misunderstanding, and limit social diversity. Research suggests that carefully managed "contact" between young people can break down barriers to communication, and that integrated education facilitates cross-communal interaction between pupils from different backgrounds (Irwin, 1993; Stringer et al., 2000).

Yet, it is not clear what it means for teachers to "do," still less, "learn to do," integrated education. Unlike other forms of multicultural education (Gillborn, 2004) it is self-evidently neither a pedagogy nor a practice. The theoretical role of pedagogy in the "contact" theory that supposedly informs integrated education is ambiguous. This suggests a need to analyze what pedagogy means to teachers and how their practice gives meaning to their educational role (Bourdieu, 1972).

This chapter begins by briefly tracing the background of integrated education in Northern Ireland. Next, it examines possible approaches to the pedagogy of integrated education. This brings together several notions: "visible" and "invisible" pedagogic discourses suggested by Bernstein (1997); the concepts of "communities of practice" or "learning community" introduced by Lave and Wenger (1999); and ideas about the epistemology of learning and pedagogy to explain the construction of teachers' educative role, as suggested by Engeström (1996) amongst others. The third section draws on this thinking to explain different patterns in teachers' understanding of integrated pedagogic discourse as revealed in two recent research reports on integrated schools in Northern Ireland. The final sections try to draw some general conclusions about the relationship between integrated education and the negotiation of meaning in pedagogic discourse.

## Integrated Education Development in Northern Ireland

Education in Northern Ireland has commonly been associated with problems of conflict and social change. Dunn and Morgan (1999) note a "continuous tension" between "models which emphasise the importance of education as a mirror which reflects the values and structure of society and helps to maintain and perpetuate them, and those which concentrate on education's potential as an agent of change" (p. 141). This view may be too simplistic; but it conveys the underlying idea of tension between the assumed either/or function of education in sustaining and maintaining social categories, or creating and maintaining new, or less prescriptive and less divisive ones.

Integrated education in Northern Ireland has been represented both as part of a reconstruction of new shared identities and part of the multicultural "Peace Process." Originally a voluntary, parent-led, Christian, interdenominational and ecumenical movement (Spencer, 1987) it was also inspired by the child-centered, deschooling movement of the 1970s. Subsequent, unanticipated formal support from the government under the Education Reform N.I. Order 1989 linked it to statutory policies, such as the "cross-curricular" themes of Education for Mutual Understanding (EMU) and Cultural Heritage (CH) for all Northern Ireland schools, as well as to British policies designed to increase parental choice and/or school competitiveness.

There are now over 50 integrated schools (including 18 secondary) comprising around 5 percent of the Northern Ireland student population. Funded by the government, the schools are seen as a separate integrated sector alongside five other sectors: the (de facto Protestant) state sector, a comparable voluntary Catholic sector, academically selective, mainly voluntary, grammar schools, and Irish language and independent Christian sectors. Integrated schools have attracted much wider attention (both hostile and supportive) than their relatively small numbers might seem to justify (Darby, 1987). The Northern Ireland Council for Integrated Education (NICIE, 2000) asserts "that children are not the problem but rather the custodians of the solution". Yet it is unclear whether this means that integrated education will lead to the kind of society in which conflict is not a problem or whether the success of integrated schools means that conflict is already becoming manageable.

This confusion of means and ends was addressed in a report published by Johnson (2002). It was argued that the pivotal role of teachers in the success of integrated schools had been neglected. While acknowledging that integrated teachers are appointed to reflect ethnoreligious balance, Johnson argued that this did not necessarily guarantee a multicultural ethos in integrated schools. She suggested that shared multicultural education training is required to prepare integrated teachers for their special task.

But ensuring that teachers can learn to "become" integrated may be more complex than this suggests. Most are currently recruited from segregated sectors and must learn "on the job." (The "Good Friday/Belfast Agreement" of 1988 requires that the Catholic/nationalist and Protestant/unionist communities must be accorded "parity of esteem," making it unlikely that existing separate communal institutions will change in the foreseeable future.) It is true that integrated teachers are mastering increasingly sophisticated multicultural class management techniques; but educational research suggests this does not necessarily mean that their basic professional or personal approach changes (Thrupp et al., 2003).

Connolly (2003), an advocate of critical multiculturalism, argues plausibly that teachers must encourage pupils to "challenge existing stereotypes and prejudices" and "develop a more grounded appreciation of their own identities and also those of others" (p. 180). But this kind of work requires a major realignment of teachers' professional perspectives and/or a highly literal and personal "transformation," which, even with the best of intentions, teachers may find difficult to achieve (Gillborn, 2004; Sleeter, 2004). The culturally loaded role of teachers, and the pedagogy of what and how they teach, is thus particularly important for understanding how integrated schools work. The question is, how do integrated teachers construct an appropriate everyday classroom practice, in what is likely to remain an otherwise largely socially and ethnoreligiously segregated society?

### Constructing Pedagogy and Teachers' Practice

One of the problems with social identity and "contact" theories which inform thinking about conflict and education in Northern Ireland (Bloomer and Weinreich, 2003) is their failure to take adequate account of the constitutive nature of educational processes (Mehan, 1996; Wells, 1999;) and the ideas that legitimize teachers' practices (Bernstein, 1999; Dussel, 2001). A "constructivist" approach can help to clarify some of these issues. Bernstein's (1997) work on social class and education distinguished analytically between two different "ideal type" or forms of pedagogic discourse: visible pedagogy—realized through a relatively restricted, explicit, language of hierarchical social control, and strong classification and framing of knowledge (discrete subject boundaries, for instance); and invisible pedagogy—realized through weak classification and "elaborate interpersonal communication" (Bernstein, 1997, p. 67). His study of the invisible, child-centered, pedagogy of infant education is a paradigm case of the latter. The infant teacher's interpretation and thus "control" of pupils' learning, is realized indirectly through theories (and an epistemology) of natural child development, and the management of physical space and surveillance for spontaneous signs of infant "readiness" (Bernstein, 1997, p. 60).

Within educative practice the notions of visible and invisible pedagogic discourse invoke different roles for epistemologies of knowledge in theories of learning. A visible pedagogy is more likely to be associated with a "transmission-acquisition" model of learning and an idea of knowledge as objective, decontextualized and canonical, or authoritative (Gorodetsky and Barak, 2004). In contrast, an invisible pedagogy is more likely to be associated with a dialogic and participative model of learning and an epistemology of knowledge as tacit, relational, and emergent (Bruner, 1996; Wells, 1999). Arguably, teachers' understanding of appropriate pedagogic practice must be a dialogic and, relatively speaking, invisible, shared model of learning—a "community of practice," in which "learning and the development of a sense of identity are inseparable" (Lave and Wenger, 1999, p. 31).

Visible and invisible pedagogies can also be mapped on to the two polar extremes of what has come to be seen as a continuum of teachers' modes of "being" professional practitioners. One end of the continuum is represented by a model of the cognitively engaged teacher, applying rationally based principles and propositional evidence continuously and routinely to action (Eraut, 2000). The other is represented by Schon's (1983) "reflective practitioner" critically deploying "reflection-in-action" to expand critical or creative self-awareness (McNiff, 1993).

The visible pedagogy of the rational teacher has been criticized for dehumanizing the learning relationship and for its realist conceptualization of knowledge and "knowing" (Elliott, 1989). An invisible pedagogy of reflective practice may also have limitations. Lawless and Roth (2001) and van Manem (1995) argue that conscious, cognitive awareness and/or critical disengagement may often be unrealistic in ordinary classroom experience, particularly for newcomers and probationers, and even for long-standing teachers, especially those who lack a range of varied teaching experience (Gorodetsky and Barak, 2004). But Lave and Wenger (1999) argue that this can be overcome by apprenticeship or internship/practicum models of teacher induction which provide "situated opportunities" to learn as a "legitimate peripheral participant" (p. 30).

Taken together these contrasting perspectives suggest an interpretive framework for assessing the extent to which integrated education pedagogy and practice realize one or other of the two ideal types. The question of interest is what "doing," integrated education actually means for teachers. For instance, does talking about integrated education amount to a visible representation of practice, of how things ought to be done? Or does talking "about" or "within" the practice of integrated education in some senses constitute the activity of integrated education itself and the means for its collective "expansive development" as suggested by Engeström (1996, p. 67)? Holquist (1996) and van Manem (1976, 1995) suggest that narrative and metaphor form a discursive context that makes the consequences of practice meaningful, enabling actions to give meaning to words, as well as the reverse. Is this Bernstein's invisible pedagogy and also the discursive context that Lave and Wenger (1999) see as especially important for "peripheral participants"? If so, how robust is it? As Lave points out elsewhere (1996, p. 24) talk as action, or, which is inconsistent with action, can sometimes cause misunderstanding and lead to retreat from interpretive, creative talk into a decontextualized, visible pedagogy of authority and control.

## Learning "to Do" Integrated Education

Whatever their complexity, the concepts of visibility and invisibility in pedagogic discourse provide a useful means of identifying how teachers understand both "doing," and "learn to do" integrated education in schools in Northern Ireland. There are strong indications that teaching in the first integrated schools was an inherently invisible pedagogy and practice. Fraser (1993) found that being a "pioneer" integrated teacher was largely "an attitude of mind" (p. 85). "Teaching was a 'practical' way of countering what was considered to be wrong with segregated education and for them . . . always better than just speaking out against it" (p. 86). However, not all the evidence points the same way. Some early teachers expected integrated schools to "feel different" and were surprised when they were not. The official NICIE guidance is also unclear about how to interpret integrated schools pedagogy (NICIE, 2000, 2001).

A detailed inspection of recent research on integrated schools reveals a complex and multilayered situation. Two recent reports, Integrated Education in Northern Ireland: Integration in Practice by Montgomery, Fraser, McGlynn, Smith and Gallagher (2003) and The Practice of Integrated Education in Northern Ireland by Johnson (2002) reflect this complexity. An interesting feature of both reports is the extent to which they judge many, if not most, integrated education teachers to be unclear about how to characterize the pedagogy of integrated education. Montgomery et al. found that integrated teachers' practice eluded definition even when it is observed "in practice" (p. 17). The teachers they interviewed emphasized integrated education's "child-centered focus," yet were reported to have had "difficulties in explaining how children can or do become integrated (Johnson, 2002, p. 24). They complained that there was no "base-line for judging levels of integration attained . . . Repeating the same terminology to explain the integrated ethos was easier for staff than finding the vocabulary to define what it might mean in practice. The ethos was interpreted differently in practice by different people. Deciding what to do and how to do it was therefore far from simple"(Johnson, 2002, p. 31).

Johnson's report reiterates these recurring themes. Teachers spoke of feeling that their classroom practice fell short of the "integrated ethos." Johnson's judgment is that they were "only confident" about promoting inclusion and tackling issues of prejudice or discrimination where there was a more "balanced" enrollment. Where there was an imbalance, there was "a purposeful norm of 'protective silence' that kept the notion of cultural identity indiscernible" (p. 11). While some teachers reported improvisation in the use of methods that supported a cultural diversity approach, such as discussion activities, debates and group methods, there was no systematic approach. "It was difficult for the individual teacher to figure out how much, what, when and in which ways such discussion is considered appropriate" (p. 23). As noted above, Johnson suggests "that some of the reticence . . . stems from the lack of formal training or preparation (teachers) felt they have . . . and because they come from their own respective traditions and typically have been educated and trained in single-identity schools themselves."

Another source of this apparent lack of confidence may be the reported spontaneous idealism of the original integrated movement (Dunn and Morgan, 1999; Morgan et al., 1992). One teacher recalled, "In the past, we used to have an evening in somebody's house and we would just throw out pieces of the newspaper, or the bible, etc. and then we would talk about what that meant to us" (Montgomery et al., p. 21). But such enthusiasm could eventually be worn down: "I came in blazing with enthusiasm . . . (but then) . . . the pressure of the curriculum and everything else takes over completely" (Montgomery et al., p. 21). The paradox, as Johnson (2002) points out, is that although initially some teachers may not have had a high commitment, they seemed—almost as a revelation—to have acquired a genuine respect for the "ethos" and its "incredibly positive influence in young people's development and confidence in their schools' success in achieving an integrated ethos" (p. 22).

The apparently spontaneous nature of the pedagogy or practice of integrated education and pioneer teachers' attitudes to it, clearly in part derives from the legacy of early joint parent-teacher mutual support networks (Morgan et al., 1993). This experience as "knowledge" about integrated practice seems to have had the informal quality of Lave and Wenger's (1999) "legitimate peripheral involvement." For early teachers, "doing integrated education" was largely a matter of "improvisational practice" informed by interaction with parents, pupils and other professionals involved in the movement. Johnson's teachers said they had "learned to count on each other (other teachers or staff) for ideas and problem-solving" (p. 22). Principals seemed to think this informal learning process is still going on. "Well, the pioneer spirit isn't here as it was in the beginning . . . (but) we're all learning about each other" (Montgomery et al., p. 19). Another principal said, "I would hate to think that you would ever have achieved integration . . . I think that is the most exciting part of this whole situation . . . it is a process, for the individual, for the school and for the movement" (Montgomery et al., p. 17).

## Tensions in the Invisible Pedagogy

There are clear indications in both research reports that the pedagogy of integrated education had the potential to engender engagement and commitment; but that its "elusive" or "invisible" character could also produce tension and a sense of frustration in everything to do with " talking about" it. For instance, regarding the problem of balance, Montgomery et al., observe, it "was not just about maintaining or trying to maintain a religious balance in pupil numbers" but "a kind of balancing act which schools felt compelled to manage in order to convince parents and the community that the school's promise to treat all pupils equally, was indeed being fulfilled" (p. 31). Worries about not having the "right approach," as Johnson also comments, could have disproportionate effects on confidence.

A sense of frustration with the fact that integrated pedagogy was rather imprecise particularly affected principals. Montgomery et al. infer that for them the process was not "automatic"; they were aware that something ". . . had to be put in place by the school . . . it's not an on-going process." Yet as one principal said, "[P]utting it into practice is quite another matter" (Montgomery et al., p. 17). One way of "making things happen" was to appoint an "integration coordinator" or begin an "action plan." But this could also lead to a paradox. While it implied action, it also seemed to expose more inaction and ineffectualness. An "integrated coordinator" said, "At the moment, we need to talk to staff and re-evaluate where they are . . . how they can progress from this point on. I am beginning to think we are not integrating as people . . . and we are not actively integrating our pupils" (p. 32).

A related frustration in the matter of child-centeredness for teachers and principals in postprimary integrated schools (not discussed in any detail by Montgomery et al.) concerned the all-ability aspect of integrated education. This was not just the problem of teaching in a way that respects both difference and inclusivity (lower-ability/SEN (Special Educational Needs) and also high-ability pupils) but also in articulating both as inherent aspects of the integrated ethos. Johnson suggests that this is a more important tension than that of achieving religious balance. She particularly links the tension over mixed ability teaching to teachers' concerns about the inadequacy of provision for meeting students' more challenging personal and emotional needs and teachers' own training needs "in areas such as cultural diversity, human relations, conflict resolution and mediation skills, behavior management, special education and all-ability teaching strategies" (p. 24).

One of the effects of such tensions was to focus attention and energy on the psychological or organizational need to make integrated education palpable or "real" in the daily life of the school. For some teachers, especially at primary level, this was apparently experienced in a highly personal way. For instance, one primary teacher said: "As a teacher, you have to start by demonstrating ways of behaving and interacting in your own practice with whomever you meet" (Montgomery et al., p. 21). Being an "integrated teacher" seemed to involve an imperative to perpetually "model" the target behavior. Yet other teachers who had worked in other sectors had a more relaxed view. Their personal style of teaching mattered, but otherwise, integrated schools were felt to be "not that different" (Montgomery et al., p. 20).

## Pressures to Make Invisible Pedagogy Visible

There seems to be little doubt that tension over what "doing" integrated education means helped to increase pressure to make it visible. Johnson suggests that such tensions can be overcome if teachers are taught how to be "integrated teachers," in particular, in terms of diversity and conflict mediation training. But although many teachers would seem to endorse this call, they appear ambivalent about whether training is the answer to all their problems. As one said, "You've either got [it] in here or you don't (pointing to heart and head); you can't really train a teacher to believe in principles of integration" (Johnson, p. 23). A not necessarily inconsistent ambivalence was noted by Montgomery amongst other teachers: "Teaching was considered rather a specific practice consisting of a series of challenges and rewards which existed irrespective of the particular sector or school they happened to be in" (p. 9).

There was, however, a concern about the need to ensure a balance between making the person-oriented, child-centered, style of teaching manifest to pupils, and at the same time demonstrating integrated education's capacity to respect different community traditions. It was evident in seemingly increasing requests for explicit guidance from school advisers. The teachers' guidelines from NICIE (2000) represented the official view that personal experience should be exemplary: Teachers, it states,

> must monitor their own behavior for instances of bias, for example, of gender, religious, political allegiance or sexual orientation . . . Teachers in an integrated setting can learn a lot from their peers about the religious, cultural and political background of their pupils. Of course they can also learn from the pupils themselves, but by learning from each other *before* they go into the classroom, teachers can avoid expressing unintentional bias or prejudice." (p. 13)

Yet Montgomery et al. found that interaction on cross-communal matters was at best only beginning to be addressed in many integrated staffrooms (p. 28).

The "integrated teacher" nevertheless, had to be constantly alert to his or her own shortcomings and be prepared for opportunities to demonstrate, model, or otherwise enable, pupils to infer meaning from experience. Prejudicial remarks or behavior in particular, it was implied, could be useful for emphasizing the phenomenological potential of experience. Thus, rather than ignoring "passing remarks," NICIE (2001) exhorts teachers to seek them out: "Integrated and mixed settings provide particular challenges for teachers: picking up small comments from pupils and bringing them into the open for discussion means confronting openly the tensions that exist in Northern Irish society" (p. 13).

This integrated strategy was not always easy according to Johnson's (2002) respondents and could lead to a form of what might be called "pedagogic retreat." She notes, "[T]eachers' concerns about offending or putting the minority at a disadvantage was [*sic*] formidable enough to restrict their involvement at this level. They only discussed issues when the need arose, e.g. when a student made a prejudicial remark" (p. 16). As one teacher commented, "I use humor to address these situations; some teachers might refer the concern after the fact to the year head for guidance; but not everyone is as comfortable or competent knowing what to do" (p. 27).

Arguably, this explains the ambivalence about the prohibition on using divisive symbols. According to Montgomery et al., "The majority of schools prohibited the wearing or display of symbols considered to be potentially divisive. These could however be used as an aid to instruction" (p. 30). Focusing deliberately on divisive symbols could be an instructional pedagogy that revealed the core mission of integrated education to bring about reconciliation; but understandably was acknowledged to be a high risk strategy. However, Johnson (2002) seems more confident than her respondents: "Some integrated schools have experienced particular difficulty in determining a fair and balanced policy regarding the display of cultural emblems/symbols; no matter how difficult it is, such policies need to be established with the input of everyone recognizing the integrated mission of the school" (p. 27).

## Discussion

These extracts do not do justice to all the points made by the two reports; but they give something of the flavor of the complexity of what "doing" and "learning to do" integrated education might mean to teachers. It is important not to get it out of proportion, however. Despite the apparent frustration, neither report suggests that integrated schools were failing their pupils. On the contrary, Montgomery et al. conclude that the vast majority, especially those with least prior diversity experience, had benefited. Pupil satisfaction was high and exposure to diversity and difference was found to have encouraged a "re-definition of self" (p. 44). Johnson also accepts that almost all teachers were confident their schools had achieved an integrated ethos and their students were much better off having experienced diversity. "We are living in an abnormal society here in Northern Ireland, but here (name of school) is a normal school where everyone is accepted."

To understand what might be happening in integrated schools in Northern Ireland and why there seem to be pressures for the initially "invisible" pedagogy of integrated education to become more "visible," it is important to appreciate what is distinctive about the knowledge and epistemology of integration for teachers. So far it has been suggested that initially integrated education was based on an epistemology that entailed implicit rather than explicit, relational rather than objective, and emergent rather than canonical knowledge, and thus required a participative and jointly created, rather than a "transmission-acquisition" model of learning. Arguably this is still what seems to be required. But this is not quite the same as saying that integrated education knowledge is part of a "hidden," "informal" or "communal" ethos or curriculum (Irwin, 1993); or that students' attitudes are influenced by cross-communal friendships, although they clearly are. It is not even that integrated schools often succeed in adopting a "whole school" approach to Education for Mutual Understanding ((NI) CCEA, 1997). It is that in integrated schools doing integrated education means creating knowledge jointly, as an invisible pedagogy or epistemology of shared practices.

The implications of this are profound for integrated schools. Another way of putting it is to say that the "knowledge" of integration has become part of the culture of the school; but that at the same time, it has also made that culture considerably more complicated. It is as if the school had literally become more like the "learning community" described by Lave and Wenger (1999) with teachers "extending what they know beyond the immediate situation" through a "dialectical process of . . . helping to produce new understanding" (Lave, 1996, p. 13). In this process, language—already so vital in education—has had to become considerably more subtle, complex and elliptical, since it is no longer primarily about affirming a single view of reality, but about exploring different worlds and, for teachers in particular, sharing or negotiating meaning around different kinds of literal and metaphorical talk about pedagogy and practice (Moffat, 2004).

The problem for integrated schools seems to arise in managing this complex process of linguistic adjustment. The evidence seems to suggest that pressure for more "visibility" and "explicitness" creates tensions and a tendency for the language of integration to assume "transmission-acquisition" characteristics that separate it from daily classroom practice and staff room "chat." Potentially, this can set up a vicious circle of frustration and demands for clarification. Problems can also build up around the pressure to make integrated education visible and palpable for parents and people outside school. Together with other pressures—such as demands for accountability and competition for pupils in an otherwise religiously and academically selective system—it has the potential to trigger a negative spiral of retreat into explicitness and defensive inflexibility.

## Conclusion

One of the lessons that can be drawn from this review of teachers' understanding of integrated pedagogy and practice is that integrated education does not seem to be a "powerful" kind of knowledge to be transmitted or taught. Nor can it easily be "decontextualized." It is true that integrated schools have an explicit ethos and curriculum that, for instance, emphasizes tolerance and mutual understanding, and rules and norms that define and constrain roles, according to the perceived ethnoreligious balance. But the evidence suggests that integration education is best understood as a kind of continuously interpreted activity in the sense defined by Lave (1996). Arguably, it also works best when staff themselves understand it as an ongoing process of participating in a "community of learning" in which everyone—students and parents too, if possible—plays a part (Tyrrell, 2002).

But it is not easy to maintain this kind of learning as a "virtuous circle," "enabling metaphor" or "shared culture" of pedagogic understanding (Fullan, 1993; Moseley, 1993; Yanow, 2000). In part, this is because it is hard to convey precisely what being an integrated teacher entails. In many ways it might be considered as only a nominal or honorary status, which permits the incumbent to talk with others *as though* his/her pedagogic activity entailed "doing" integrated education. The reports discussed here seem to confirm that teachers often only learn the consequences of practice by sharing interpretations with colleagues and acquaintances. As Lave and Wenger (1999) make clear, practitioners' talk "about" and "within" practice can have many quite different functions: "engaging, focusing, and shifting attention; bringing about coordination, etc. on the one hand; and supporting communal forms of memory and reflection, as well as signaling membership on the other" (p. 30). This is an important challenge for the induction of new or "learning" integrated teachers; for them, "the purpose is not to learn from talk as a substitute for peripheral participation; it is to learn to talk as a key to legitimate peripheral participation" (Lave and Wenger, 1999, p. 30).

A second, general, though more tentative lesson from this analysis, is that much of the "knowledge" or understanding about integrated education can only come from others, or as Levinas might argue, the capitalized, singular "Other" (Standish, 2004). Such learning (not necessarily from teachers) requires a willingness to be open to experience, or at least to acknowledge the contingency of one's own experience and that of others—and in ways which may not be readily represented in discourses that typically legitimate educational practices. Talking about integrated education may be a way of becoming a "legitimate peripheral participant." The next step lies in moving beyond this kind of talk to creative interpretation and negotiation of shared understandings of pedagogy and social practice.

## References

Bernstein, B. (1999). Vertical and horizontal discourse: An essay. *British Journal of Sociology of Education*, 20 (2): 157–175.

———. (1997). Class and pedagogies: Visible and invisible. In Halsey, A.H., Lauder, P., Brown, A.S. and Wells, G. (Eds.), *Education, culture, economy and society* (pp. 59–79). England: Oxford University Press.

Bloomer, F. and Weinreich, P. (2003). Cross-community relations projects and interdependent identities. In Hargie, O. and Dickson, D. (Eds.), *Researching the troubles: Social science perspectives on the Northern Ireland conflict* (pp. 141–162) Edinburgh: Mainstream.

Bourdieu, P. (1972). *Outline of a theory of practice*. Cambridge: Cambridge University Press.

Bruner, J. (1996). *The culture of education*. Cambridge, MA: Harvard University Press.

Connolly, P. (2003). The development of young children's ethnic identities: Implications for early years practice. In Vincent, C. (Ed.), *Social justice, education and identity* (pp 166–184). London: RoutledgeFalmer.

Darby, J. (1987). *Managing conflict*. London: Minority Rights Group Report

Dunn, S. and Morgan, V. (1999). A fraught path—Education as a basis for developing improved community relations in Northern Ireland. *Oxford Review of Education*, 25 (1): 141–151.

Dussel, I. (2001). What can multiculturalism tell us about difference? The reception of multicultural discourses in France and Argentina. In Grant, C.A. and Lei, J.L. (Eds.), *Global constructions of multicultural education: Theories and realities* (pp. 93–111). New York: Erlbaum.

Elliott, J. (1989). Educational theory and the professional learning of teachers: An overview. *Cambridge Journal of Education*, 19 (1): 81–101.

Engeström, Y. (1996). Developmental studies of workers as a test bench of activity theory: The case of primary medical practice. In Chailkin, S. and Lave, J. (Eds.), *Understanding*

*practice: Perspectives on activity and context* (pp. 64–103). Cambridge: Cambridge University Press.
Eraut, M. (2000). Non-formal learning, implicit learning and tacit knowledge in professional work. In Coffield, F. (Ed.), *The necessity of informal learning* (pp. 12–28) Bristol: The Policy Press.
Fraser, G. (1993). Teaching in integrated schools. In Moffat, C. (Ed.), *Education together for a change: Integrated education and community relations in Northern Ireland* (pp. 85–95). Belfast: Fortnight Educational Trust.
Fullan, M. (1993). *Change forces: Probing the depth of educational reform.* London and Philadelphia: Falmer.
Gillborn, D. (2004). Anti-racism: From policy to praxis. In Ladson-Billings, G. and Gillborn, D. (Eds.), *The RoutledgeFalmer reader in multi-cultural education* (pp. 35–48). London: RoutledgeFalmer.
Gorodetsky, M. and Barak, J. (2004). Extending teachers' professional speilraum: co-participation in different experiential habitats. *Reflective Practice,* 5 (2): 265–281.
Holquist, M. (1997). The politics of representation. In Cole, M., Engestrom, Y., and Vasquez, O. (Eds.), *Mind culture and activity : Seminal papers from the laboratory of comparative human cognition* (pp. 389–408). Cambridge: Cambridge University Press.
Irwin, C. (1993). Making integrated education work for Pupils. In C. Moffat (Ed.), *Education together for a change: Integrated education and community relations* (pp. 68–84). Belfast: Fortnight Educational Trust.
Johnson, L.S. (2002). The practice of integrated education in Northern Ireland: The teacher's perspective. UN/INCORE Research monograph series. Retrieved on May 20, 2004 from www.incore.ulst.ac.uk/home/publication/occasional/
Lave, J. and Wenger, E. (1999) Learning and pedagogy in communities of practice. In Leach, J. and Moon, B. (Eds.), *Learners and pedagogy* (pp. 21–33). Open University.
Lave, J. (1996). The practice of learning. In Chailkin, S. and Lave, J. (Eds.), *Understanding practice: Perspectives on activity and context* (pp. 3–32). Cambridge: Cambridge University Press.
Lawless, D.V. and Roth, W. (2001). The spiel on "spielraum and teaching" *Curriculum Inquiry,* 31 (2): 229–235.
McNiff, J. (1993). *Teaching As Learning: An Action Research Approach.* London: Routledge.
Mehan, H. (1996). Beneath the skin and between the ears: A case study in the politics of representation. In Chailkin, S. and Lave, J. (Eds.), *Understanding practice: Perspectives on activity and context* (pp. 241–268). Cambridge: Cambridge University Press.
Moffat, C. (2004). Learning peace talk in Northern Ireland: Peer mediation and some conceptual issues concerning experiential social education. *Pastoral Care in Education,* 22 (4): 13–21.
Montgomery, A., Fraser, G., McGlynn, C., Smith, A. and Gallagher, A. (2003). *Integrated education in Northern Ireland.* Coleraine: UNESCO Centre, University of Ulster.
Morgan, V., Fraser, G., Dunn, S., Cairns, E. (1992). Views from outside—Other professionals' views of the religiously integrated schools in Northern Ireland. *British Journal of Religious Education,* 14 (3): 169–177.
———. (1993). A new order of co-operation and involvement? Relationships between parents and teachers in the integrated schools. *Educational Review,* 45 (1): 42–52
Moseley, J. (1993). *Turn your school around: A circle time approach to the development of self.* England: Wisbeck Learning Development Aids.
(NI)CCEA (1989). *Mutual understanding and cultural heritage: Cross-curricular guidance materials.* Belfast: Northern Ireland Council for Curriculum, Examinations and Assessment.
NICIE (2001). *What's what in integrated education.* Northern Ireland Council of Integrated Education.
NICIE (2000). *Integrating through understanding.* Northern Ireland Council for Integrated Education.
Schon, D. (1983). *The reflective practitioner: How professionals think in action.* London: Temple Smith.

Sleeter, C. (2004). How white teachers construct race. In Ladson-Billings, G. and Gillborn, D. (Eds.), *The RoutledgeFalmer reader in multi-cultural education* (pp. 263–278). London: RoutledgeFalmer.
Spencer, A.E.C.W. (1987). Arguments for an integrated school system. In Osborne, R.D., Cormack, R.J. and Miller, R.L. (Eds.), *Education and policy in Northern Ireland* (pp. 99–113). London: Policy Research Institute.
Standish, P. (2004) Europe's continental philosophy and the philosophy of education. *Comparative Education*, 40 (4): 486–500.
Stringer, M., Wilson, W., Irwing, P., Giles, M., McClenaghan, C. and Curtis, L. (2000). *The impact of schooling on the social attitudes of children*. Belfast: The Integrated Education Fund.
Thrupp, M., Mansell, H., Hawksworth, L., and Harold, B. (2003). Schools can make a difference—But do teachers and heads and governors really agree? *Oxford Review of Education*, 29 (4): 471–484.
Tyrrell, J. (2002). *Peer mediation: A process for primary schools*. London: Souvenir Press.
UN CESCR (United Nations Committee on Economic, Social and Cultural Rights) (2002). Consideration of reports submitted by States parties under Articles 16 and 17: Concluding observations of the committee on compliance with the convention, United Kingdom and Northern Ireland. E/c.12//1/Add.79: Recommendation 42. Geneva: United Nations.
van Manem, M. (1995). On the epistemology of reflective practice. *Teachers and Teaching: Theory and Practice*, 1 (1): 33–50.
Wells, G. (1999). *Dialogic inquiry: Towards a socio-cultural practice and theory of education*. Cambridge: Cambridge University Press.
Yanow, D. (2000). Seeing organizational learning: A "cultural" view. *Organization*, 7 (2): 247–268.

# Chapter Twelve
# From War to Peace: An Analysis of Peace Efforts in the Ife/Modakeke Community of Nigeria

*Francisca Aladejana*

### Introduction

A community has moved from war to peace. Presently, the community is enjoying some state of relative peace and life seems to have returned to normal. The various amounts of reconstruction and reunion taking place led credence to the fact that indeed there is restoration of peace. The culture of peace is becoming entrenched again. That is the culture of peace defined by the United Nations as

> a set of values, attitudes, modes of behavior and ways of life that reject violence and prevent conflicts by tackling their root causes to solve problems through dialogue and negotiation among individuals, groups and nations.

The community is the Ife/Modakeke community of Nigeria. Nigeria is a republic in West Africa and one of the most highly populated countries in the subregion, with approximately 127 million people (Federal Ministry of Information and Cultural Affairs, 2002).

Davies (2004), in her analysis of the roots of conflicts identified pluralism or diversity in terms of ethnicity, religion, tribalism, and nationalism as one of the antecedents of conflicts. According to her, while pluralism can be positive, a large number of armed conflicts are those defined by ethnic or other forms of difference. Pluralism is a factor in Nigeria which has about 250 ethnic groups which gives the country a rich culture, but this also poses major challenges to nation building. The ethnicity is so varied that there seems to be no definition of a Nigerian beyond that of someone who lives within the borders of the country. The three largest ethnic groups in Nigeria are the Hausa-Fulani, Yoruba, and Igbo. Other ethnic groups include the Kanuri (in Borno), Tiv (in Benue), Ibibio and Efik (Calabar), Edo (Benin region), and the Nupe (in Bida area). Although small by Nigerian standards, each of these lesser ethnic groups has more members than almost any other African ethnic group (Ministry of Information and Cultural Affairs, 2002).

The country's official language is English but there is a great diversity in the languages with about 400 identified ones. The most common ones are Hausa, Yoruba, and Ibo. Others are Fulfulde, Kanuri, Ibibio, Tiv, Efik, Edo, Ijaw, and Nupe. These native languages have several distinct regional dialects, sometimes with linguistic variations across short distances. There is also the Pidgin English which is English combined with native languages. In terms of religion, Christianity, Islam, and indigenous religions are central to how Nigerians identify themselves. According to the 1991 census, 47 percent of Nigerians were Muslims and 35 percent Christians.

## Conflicts in Nigeria

A number of ethnic clashes and violence have rocked the very existence of Nigeria as a nation. Amongst these are the riots between the Hausa and Yoruba in various parts of the country (Kano/Kaduna in 2000; Shagamu in 1999; Ketu-Mile 12 in 1993; Bodija in 1999 and Hausa-Igbo in 1966). There have also been some interethnic conflicts, for example, between the Ijaw-Ilaje of Ondo State in 1999, Ijaw-Itsekiri of Delta State in 1999, Aguleri-Amuleri in Anambra State in 1999, Zagon-Kataf of Kaduna state in 2000, and the Ijaw-Urhobo of Delta State in 2001. The Ife-Modakeke community indigenes are Yorubas of the South-West zone of Nigeria and the crisis between them dates back to 1898. Since that period till 2002, the communities have gone through several communal wars. According to Aladejana and Aladejana (2003), the Ife-Modakeke conflicts were as a result of unemployment, lack of education, land issues, discrimination, subjugation, flagrant disobedience to cultural values, and lack of awareness of the interdependence of people.

During these conflicts, youths (most of who are of secondary school age) were mostly involved as soldiers/fighters. Consequently, the conflicts push the most active, volatile, and most vulnerable segment of the population into antisocial behaviors and violence. The Nigerian youths are mobilized and their nationalistic and patriotic values manipulatively turned into ethnic militias (Ogunbi, 1987). During these crises, the children, the youths, and women were always on the receiving end as the children and youths became orphans and the women became single parents (Aladejana, 2004). One of the major factors, which can facilitate the involvement of youths in conflicts and make them vulnerable to violence, is a social environment that is polluted with antiethnic values, particularly when the youths are directly or indirectly exposed to social ills of dishonesty, corruption, greed, selfishness, fraud, and rebellious attitude (Ogunbi, 1987). Politics can only keep these youths out of war, but it is education that can change all these social ills and make for a lasting peace.

## Peace Education and Social Studies

Peace attitude often comes slowly and there is no one single model of achieving this. Galtung (2003) recognized the need to foster a culture of peace through education. In explaining the contribution of education as being active in the struggle for peace, Davies (2004) came up with the complexity theory. One of the features of this theory is the nonlinearity of education that refers to the phenomenon that components of a system are interdependent, so that one part intricately and unpredictably affects the operation of another. Thus, systems such as schools within the community and community within the wider society are interdependent. The Nigerian National Policy on Education (2004), in expressing possible ways of averting conflict and achieving peace stated that not only is education the greatest force that can be used to bring about redress, but is also the greatest investment that the nation can make for the quick development of its economic, political, sociological, and human resources. This position is in line with the United Nations recommendation of fostering a culture of peace through education (United Nations, 2000).

There is therefore the need for peace education. According to Lasonen (2004), peace implies an absence of conflict and quarrel, the enjoyment of security, well-being, and harmony. Any education that will provide an individual with the capability to attain peace for themselves and others is peace education. This is positive peace that is reflected not only in the absence of violence, but also in the attitude of the individuals and groups. According to Fountain (1999),

> peace education refers to the process of promoting the knowledge, skills, attitude and values needed to bring about behavior changes that will enable children, youth and adults to prevent conflict and violence, both overt and structural; to resolve conflict peacefully; and to create conditions conducive to peace, whether at an intrapersonal, interpersonal, intergroup, national and international level.

The Nigerian government has risen to the challenge of addressing the problem of peace education in various ways although under various contexts. Some of these are the establishment of the Citizenship and Leadership Training Centre in 1980 and the introduction in 1971 of Social Studies into the curriculum of primary and junior secondary schools. There was however some effort at teaching Social Studies by the Comparative Education Study and Adaptation Centre (CESAC) in 1968–79. The Nigerian Education Research Council (NERDC) in 1971, saddled with the responsibility of designing the curriculum for primary and secondary education, defined Social Studies as "a way of life, a way of seeing, viewing, conceptualizing and appreciating things, issues with special regards to their proper place and function in the reordering and management of man's total natural, social, political and cultural environment." According to Kissock (1981) and Adedoyin (1984), Social Studies is a program of study that a society uses to instill into its students the knowledge, attitude, and actions it considers important concerning the relations they have with each other, their world, and themselves.

The teachers become very important in this process as they are the most effective agents of change that perform mediating roles in citizenship issues. It is widely accepted that no education can rise beyond the quality of its teachers. Thus, how adequately trained the teachers are will determine to a large extent their effectiveness in implementing the curriculum. Their strategies of teaching are equally important. According to Nwoji-Okeke (1999), strategies represent a diverse array of activities that serve to focus attention; minimize anxiety and maintain motivation; and service to monitor learning and organize information. There is the need for teachers to devise strategies for using the available learning resources. Key strategies identified as very useful in peace education include experiential learning that builds largely on the learners' experience and participatory methodologies involving the use of role plays, simulation, case studies, television, and debates along with the regular lecture method (Bar-Tal, 2002).

## Theoretical Perspectives

This study is grounded on several theoretical frameworks. One is that of integrative peace education across the school curriculum. Integrative education has been defined as that which cuts across subject-matter lines, bringing together various aspects of the curriculum into meaningful association to focus upon broad areas of study (Walker, 1995). It reflects the interdependent real world and provides a holistic context for learning that leads to a greater ability to make and remember connections and to solve problems (Kovalik and Olsen, 1994). According to Braunger and Hart-Landsberg (1993), integrative curriculum attempts to make learning more natural and links subjects together.

The approach in this study is based on the dominant assumption that peace education must combine knowledge, skills, and attitude. In this regard, the central themes are stewardship, citizenship, and intergroup relationships with the ultimate aim of addressing both overt and structural violence in the society. Classroom practice and instructional process are essential (Readon, 1998; Fountain, 1999).

Another framework is the two-force theory by Nagler (1999) that identifies two kinds of peacekeeping: one that is based on violence, threats of punishment and the other based on nonviolent peacemaking carried out not only without the use of weapons but also in an ideal sense without reliance on coercion of any kind. Nagler recommended the latter that involves making peace through systematic, institutional use of the power of and acts of love.

A further theoretical perspective for this study is the functionalist or modernization theory that holds that educational institutions are secular national institutions that serve economic growth and political stability by building citizenship across ethnic particularisms (Mickelson, Nkomo, and Smith, 2001). The pedagogical model underpinning the introduction of the peace education in this work focuses on the content, input, and process. Thus, the curriculum content, the instructional activities required, and the teaching strategies for the implementation of the curriculum are considered.

The objectives of this study are therefore to

1. examine the extent to which the curriculum content of the JSS Social Studies reflect peace education,
2. assess the teaching strategies used to teach Social Studies,
3. verify whether or not social studies teachers are adequately equipped to teach peace education and foster a culture of peace,
4. assess if the culture of peace is reflected in the ethos and practice of the schools,
5. determine whether or not students may have imbibed the culture of peace from the learning of Social Studies, and
6. identify other factors that are responsible for the present state of peace in the area.

## Methodology

The design of the study is descriptive survey. The study looked at the formal and informal aspects of education that have been manipulated to achieve peace. Three self-designed instruments were used in gathering data for the study. One is a ten-item structured teachers' questionnaire designed to elicit information on the various strategies teachers adopt in teaching social studies. This was administered to 20 randomly selected Social Studies teachers in schools located in the conflict area under study. The second is the students' questionnaire that was divided into two sections. Section A was designed to collect information on the various activities of the school that can help to develop a culture of peace, while section B was to assess the impact of the knowledge of the culture of peace on students' attitude to peace. This was administered to 152 senior secondary school students selected by simple random sampling from three secondary schools each from the Ife and Modakeke parts of the community and 65 randomly selected Social Studies undergraduates of Obafemi Awolowo University, Ile-Ife. All these students have passed through nine-years compulsory study of primary and junior secondary school Social Studies and have been resident in the conflict area for over nine years.

The third instrument, a 20-item multiple choice test (the Basic Knowledge Assessment Test, BKAT) based on the students' study of social studies was used to assess the selected students' knowledge of key facts, concepts, and general ideas of peace and culture. No questions with difficulty level outside the acceptable range of 30–70 percent were used. The three instruments were validated using the test-retest technique and the calculated values of 0.82 (teachers' questionnaire); 0.79 (students questionnaire), and 0.83 (BKAT) were obtained.

In addition, the records of the peace efforts of the government, particularly through the Peace Committee on the Ile-Ife/Modakeke conflicts were collected through the State Ministry of Youths, Sports, and Culture. The records were used to identify the various peace efforts and their implementation that could possibly have facilitated the current state of peace in the area. Furthermore, the curriculum contents of the Nigerian JSS social studies and the teacher education program for Social Studies at Obafemi Awolowo University, Ile-Ife, were examined and analyzed to determine their inclusion of topics relevant to peace education. These are the topics that can provide individuals with the capacity to attain stable peace for themselves and others. The numbers of such topics was counted and their percentages relative to the total number of topics were calculated.

## Results

Findings from the study showed that the curriculum of social studies contained 25.81 percent of topics related to peace education in JSS I, 33.33 percent in JSS II and 23.81 percent in JSS III. On the whole, the curriculum has 28.81 percent content of peace education topics. In JSS I, the related topics include social environment; cooperation and conflict responsibility; civil right; civil obligation; nonfulfillment of obligation, and culture. In JSS II, there are topics such as cultural patterns and historical origin, social and cultural integration, sociocultural aspect of change, and culture. Also in JSS III, among the related topics are races and racism, socialization, cooperation, and conflict.

Strategies identified for the teaching of social studies in the JSS include excursions to museums, shrines, or some local activities (15–20 percent); use of a variety of learning activities or event to convey concepts such as lecture method (100 percent); discussion (60 percent); role play (40 percent ); drama (35 percent); and debate (25 percent). Other strategies identified include the use of posters (60 percent); poems and songs (20 percent), the media (15 percent); and invitation of resource personnel (10 percent). A total of 75 percent of these teachers agreed that there are adequate textbooks (see table 12.1).

**Table 12.1**  Percentage Response on Assessment of Teaching Strategies

| No. | Items | Yes No. | % | No No. | % |
|---|---|---|---|---|---|
| 1. | Social Studies teachers are involved in planning and reviewing of the curriculum | 5 | 25.0 | 15 | 75.0 |
| 2. | Resource personnel are usually invited to teach areas where the teacher is not very conversant | 2 | 10.0 | 18 | 90.00 |
| 3. | There is the provision of local resource materials like charts, real objects for teaching | 12 | 60.0 | 8 | 40.0 |
| 4. | School/community relations are encouraged through excursions to | | | | |
| | (a) Museums | 4 | 20.0 | 16 | 80.0 |
| | (b) Shrines | 3 | 15.0 | 17 | 85.0 |
| | (c) Local activities (e.g. festivals) | 4 | 20.0 | 16 | 80.0 |
| 5. | Organize learning activities on social studies concepts with varieties of events like | | | | |
| | (a) lecture in classroom | 20 | 100 | — | 0.0 |
| | (b) role play | 8 | 40.0 | 12 | 60.0 |
| | (c) drama | 7 | 35.0 | 13 | 65.0 |
| | (d) debate | 5 | 25.0 | 15 | 75.0 |
| | (e) quiz competition | 6 | 30.0 | 14 | 70.0 |
| | (f) Competitive games/sports | 15 | 75.0 | 5 | 25.0 |
| | (g) Discussion | 12 | 60.0 | 8 | 40.0 |
| 6. | Develop poems, songs on culture and occupation of Nigerians in general | 4 | 20.0 | 16 | 80.0 |
| 7. | Develop and use video-tapes, films, television /radio programs on the social studies concepts | 3 | 15.0 | 17 | 85.0 |
| 8. | Use posters to communicate the ideals of peace and intercultural education and other concepts | 12 | 60.0 | 8 | 40.0 |
| 9. | Use of adequate textbooks from which children can cultivate a culture of peace | 15 | 75.0 | 5 | 25.0 |
| 10. | I teach social studies in a way whereby my students are actively involved in the learning activities | 6 | 30.0 | 14 | 70.0 |

It was also found that 60 percent of the teachers have adequate local resources (such as charts and real objects) for teaching. However, only 30 percent of the teachers agreed that their strategy for teaching involves students being actively involved in the learning activities and 25 percent indicated that they are involved in the planning and reviewing of the curriculum.

An analysis of the BEd Social Studies curriculum showed that in Part I, 16.67 percent of the courses offered are related to peace education, 35.29 percent in Part II, 12.50 percent in Part III and 23.08 percent in Part IV. On the whole, 22.41 percent of the courses are related to peace education. These courses include IED 153–The school and the society; IED 154–The family in traditional and contemporary societies; IED 251–Social interaction in Nigeria; IED 253–Sociopolitical structure and institutions; IED 254–Collective Behavior in Social groups; IED 258–Culture and social stability; IED 353–Community organization and development; IED 354–Leadership, Followership, and Nationalism in Nigeria; and IED 458–International and multidimensional interactions.

The results of the Basic Knowledge Assessment Test of facts, concepts, principles, and generalizations of the knowledge of peace showed that 30.88 percent of the students scored 70 and above, while 49.77 percent scored between 50 and 69. On the whole, 80.65 percent passed the test scoring above 50 percent.

The survey of the ethos and practice of school showed that there are some activities of the school that reflect the culture of peace. These include quiz competition between students of the different communities (20.14 percent); peace talks (58.26 percent); discouraging any form of arguments on issues of the war (83.48 percent); and not allowing any act of violence (96.22 percent). Some activities such as the prohibition of membership in secret cults and militias, use of drugs, carrying of weapons in school were being carried out in all the schools, (having 100 percent positive response). Also all schools had counseling services. However, student admission is still clearly along ethnic boundaries (64.15 percent) and most students do not come together to celebrate national events (65.62 percent) (see table 12.2).

It was found that in line with what students learn in their social studies classes about accommodation of other religions and cultures, some of them would now want to marry outside their own community (45.16 percent ) and see all human beings as the same in spite of differences in religion and culture (48.38 percent).

Many of the students agreed that people can live peacefully together in spite of differences in religion (55.61percent) or ethnicity (55.76 percent); that conflicts need not be settled by violence (62.82 percent) or war (66.54 percent), but rather by negotiations (55.30 percent) and communication (59.91 percent). Also, some of the students have imbibed the opinion that war is disastrous (66.82 percent) and that for peace to reign in a place, there is the need for equal rights (56.68 percent), fulfillment of mutual obligations (73.73 percent), and tolerance (50.23 percent). A total of 65.44 percent of the students are ready to change their hatred for the other warring communities.

The records of the State Ministry of Youths, Sports and Culture and the government-established Ife Peace Action Committee revealed that there have been strategic responses put in place by government to ensure lasting peace. These efforts include

1. fair allocation of state resources to cater for the community that felt cheated in the earlier allocations,
2. meeting the special requests of the two communities e.g. establishment of the local government area office in Modakeke community for which this group has agitated for a long period,
3. providing a forum for discussions and negotiations between the two groups,
4. firmness in allowing the two sides to know where they went wrong and giving room to correct such mistakes,
5. not allowing any group to feel marginalized or oppressed,
6. providing jobs for the youths who served in the militias (loans were provided for many of them to start businesses especially transport business),
7. regular radio and television programs on the bad effects of war,
8. jingles and posters that can enhance peace,
9. encouraging intermarriage between the two groups,
10. entrusting community leaders and paramount rulers with responsibility and authority to ensure peace in their areas, and
11. establishing joint forum between the different religious leaders where they can pray together and also foster some form of peaceful corelations among their various followers.

**Table 12.2** Imbibing the Culture of Peace from Social Studies

| No. | Items | Agree No. | % | Disagree No. | % |
|---|---|---|---|---|---|
| 1. | All human beings are the same in spite of differences in religion and culture. | 105 | 48.38 | 112 | 51.62 |
| 2. | Will you like to work with a non-Yoruba person | 107 | 49.31 | 110 | 50.69 |
| 3. | It is good to marry out of one's sub-ethnic/ethnic group | 98 | 45.16 | 119 | 54.84 |
| 4. | Two people of different religions can work or live peacefully together | 112 | 51.61 | 105 | 48.39 |
| 5. | A multi-ethnic society cannot live at peace | 121 | 55.76 | 96 | 44.24 |
| 6. | One is more responsible to one's ethnic group than to the country | 194 | 89.40 | 23 | 10.60 |
| 7. | Tolerance is important among people | 109 | 50.23 | 108 | 49.77 |
| 8. | War is the best way to settle major conflicts | 116 | 53.46 | 101 | 66.54 |
| 9. | Negotiations should always take place when problems arise | 120 | 55.30 | 97 | 44.70 |
| 10. | Violence is necessary at times to defend one's position | 81 | 37.33 | 136 | 62.67 |
| 11. | From history, wars are disastrous, bringing death, loss of property and life and poverty. | 145 | 66.82 | 72 | 33.18 |

**Table 12.2** Continued

| No. | Items | Agree No. | % | Disagree No. | % |
|---|---|---|---|---|---|
| 12. | There must be communication of information to avoid conflicts. | 130 | 59.91 | 87 | 40.09 |
| 13. | Fair allocation of resources is important for peace | 123 | 56.68 | 94 | 43.32 |
| 14. | It is important for mutual obligations to be fulfilled to have peace | 160 | 73.73 | 57 | 26.27 |
| 15. | I am now ready to change my hatred for the other group with which we have conflict. | 142 | 65.44 | 75 | 34.56 |

## Discussion

Social studies curriculum in the Nigerian JSS can be said to contain adequate peace education content. This is based on the premise that the subject is expected to be an integrated study of the social sciences like geography, political science, economics, sociology, psychology, and anthropological humanities in order to promote civic competency. Thus, in the present content, it can be said that peace education has a good share of the curriculum content and that it reflects the content of a good peace education program as identified by Fountain (1999).

In spite of the fact that it is the teachers who implement the curriculum, it has been found that they are not generally involved in its planning and review. One would expect them to be more involved in the planning and review of the curriculum, as they are the ones dealing directly with the learners and can thus identify the special needs and areas of deficiency in the curriculum. Most teachers have adequate local resource materials for teaching in the form of charts, posters, and real objects. This might not be unconnected with the fact the curriculum is closely related to the local environment.

The study showed that generally most schools have adequate textbooks for teaching social studies. In almost all the states of the country, the Ministry of Education recommends standard textbooks and they are made compulsory for use. The most common strategy adopted by teachers is the lecture method. Others are sports/games competition, discussion, role play, and the use of posters. The remaining strategies such as excursions, debate, drama, quiz competition, and the media are employed by only a few of the teachers. It is not surprising that very few teachers employ the media because most schools cannot afford such facilities, and also there is no electricity in many of the schools. Teaching strategies absent include experiential learning and participatory methodologies as recommended by Bar-Tal (2002) for a good peace education. Many of the learners may therefore not be adequately involved in the learning activities and many others may also not really see the link between what is being studied in the classroom and what is happening in real life outside the classroom.

The BEd Social Studies program is offered as preservice training for student-teachers who will go out on graduation to teach JSS social studies. These undergraduates take education courses and core courses (that deal with their subject of specialization). On the whole, about a quarter of the core courses are related to peace education. This can be considered adequate in view of the fact that social studies is an integrated subject that is expected to equip the teacher to teach the various topics integrated into the subject. This amount of content can equip the teacher to teach peace education and foster a culture of peace.

Students who have spent nine years in studying Social Studies in the primary and junior secondary schools are expected to have adequate basic knowledge of issues relating to peace and intercultural education. The study has shown that the exposure of the students to integrative curriculum of peace education embedded in social studies has achieved to some extent this objective. It was observed that many of the students recognize the dangers of conflict and war, are ready to accommodate others in spite of religious, ethnic, and cultural differences and would prefer negotiations and communication to settle conflicts. Many of them also recognize the need to have equal rights and the fulfillment of mutual obligations. Students have thus acquired some knowledge, skills, and attitude that can engender peace and help to resolve conflicts. This is in consonance with the ultimate aim of integrative peace education identified by Adedoyin (1984) and Braunger and Hart-Landsberg (1993).

The ethos and practice of the schools reflect the culture of peace and has probably helped to build the culture of peace in the students. Some of these school practices are quiz competitions, avoidance of any form of violence, peace talks, and counseling services. The schools are functioning as regions of peace as the climate models peaceful behavior in the relationship between staff and students. There is equality and nondiscrimination, conflicts are handled in a nonviolent manner, and there is forum for the discussion of peace and justice. All of these are in line with the ideals of a good peace education as identified by Fountain (1999). It is however found that the two communities still have a great deal of work to do to engender trust in each other. For instance, most of the schools have few students from their warring ethnic groups and students from both areas are just starting to come together for local/state/national celebrations. This is in line with Nagler's (1999) recommendation of achieving peace though through nonviolent means.

Various measures have been put in place by the government not only to achieve a state of no war, but that of real stability, confidence, and peace. These activities include : forums for discussion and negotiation, not allowing any group to be marginalized, fairness and equity in the allocation of resources, alleviation of poverty, gainful employment of the youths, involvement of paramount rulers and community leaders in peace efforts, and positive use of the mass media for peace initiatives. These activities of the government may have contributed immensely to the present state of peace.

Peace attitude and indeed imbibing a culture of peace often comes slowly. The results showed that the peace that Ife/Modakeke area now enjoys, and the culture of peace that the students are now imbibing seems to be from the study of social studies, the reflections of peace in the ethos, and practice of the schools and other government activities. They have not just helped to resolve the conflict but have improved interaction and cooperation among the warring factions.

This study has potential relevance for other contexts that are striving to achieve peace or want to sustain their current peace. There is the need to have education for peace. This can be integrated into the present subjects in the curriculum. The content should be adequate, with appropriate instructional activities included. Key strategies for implementation should be beneficial to the learners in imbibing a culture of peace. Peace will reign when all those involved can come together to the discussion table, when there is a party that is neutral enough to broker peace, when none of the groups is marginalized, when special requests can be ironed out in the interest of both parties and when there is the appropriate education both at school and out of school. A great deal of responsibility rests on the government to show total commitment to attaining lasting peace through its various activities and the provision of the required support for schools to effectively carry out peace education.

## Conclusion

Multiethnic communities that lack religious tolerance are highly susceptible to conflicts and violence. When conflicts arise, efforts must be made to stop the violence and engender a culture of peace. The activities of the government toward peace have contributed immensely to the current state of peace enjoyed in the area under study. It is important for the source of the problem to be identified and objectively examined, and possibly to have good neutral parties to broker peace. Peace can also be achieved when community leaders are involved in peace initiatives and none of the parties in the conflict is marginalized.

Education however remains the best tool to achieve lasting positive peace. Social studies, whose primary objective is to develop good citizenship and national unity, has been found to contain adequate number of topics related to peace education to serve this purpose of integrative peace education. The curriculum of the related teacher education in Nigeria also contains adequate number of topics to equip the teacher to teach the students effectively; this is very important as teachers are the change agents. Some factors such as the possession of adequate local resources and textbooks and the use of appropriate teaching strategies are found to be important to teaching the subject to achieve the desired results. It is of equal importance that the culture of peace be reflected in the ethos and practice of the school.

## Recommendations

1. Social Studies is an integrated subject that can be used to teach peace education. It should however not just be taught just as a school subject which aims at "getting knowledge," but taught with those strategies that will enable learners to imbibe the skills of anger management, impulse control, problem solving, positive thinking, compassion, learning to live together, and respect for human dignity.
2. Any peace education studies must have an appropriate curriculum content that meets the need of the area undergoing conflict.
3. The communities in other conflict areas have a lesson to learn from the various government activities that have been mobilised to ensure that this region returns to a stable state of peace. Governments must ensure that all necessary steps are taken to ensure peace.
4. It is important that research is carried out on the ways various warring communities have achieved peace to serve as point of reference for others.

## Note

I acknowledge the great efforts of my seven final year BEd (Social Studies) students, the 2003/2004 set who assisted in collecting the data for this study. They are Titilope Fatoba, Sola Sasere, Gafaru Ibrahim, Adebayo Fasanu, Jimoh Seidu, Oluyemi Olufunwa and Toyin Idowu.

I thank Clarke-Hababi, Associate Director, EFP-International and Naghmeh Sobhani, Director, EFP-Balkans, for their outstanding contributions to the EFP Program. Special thanks are due to Kimberley Syphrett and Gabriel Power for their excellent research work and to Christine Zerbinis and Stacey Makortoff for their invaluable editorial assistance.

## References

Adedoyin, F.A. (1984). Social studies and dynamic classroom interaction. *CESAC Occasional Paper*, 5. Lagos: University of Lagos.

Aladejana, A.I. and Aladejana, F.O. (2003 June). An analytical review of conflicts in Nigeria: A proposal for a culture of peace. Paper presented at UNESCO Conference on Intercultural Education, Jyvaskyla, Finland.

Aladejana, F. (2004). Programmes for fostering a culture of peace. In Lasonen, J. (Ed.), *Cultures of peace: From words to deeds. The Espoo Seminar Proceedings June 13–14, 2003* (p. 101). University of Jyvaskyla: Jyvaskyla, University Printing Press.

Bar-Tal, D. (2002). The elusive nature of peace education. In G. Solomon and B. Nevo (Eds.), *Peace education: The concept, principals and practices around the world* (pp. 27–36). New Jersey: Larry Erlbaum Associates.

Braunger, J. and Hart-Landsberg, S. (1993). Crossing boundaries: Explorations in integrative curriculum. *Database Search Detail*, ED 370239.

Davies, L. (2004). Conflict and chaos: War and education. Website of the Westmorland General Meeting Preparing for Peace" Initiative. Retrieved from http://www.preparingforpeace.org/davies.htm

Federal Republic of Nigeria (2004). *National policy on education*. Lagos: NERDC Press.

Federal Ministry of Information of Cultural Affairs (2002). *History of Nigeria*. Lagos: Federal Government Press.

Fountain, S. (1999). *Peace education in UNICEF*. New York: UNICEF Programme Division. Retrieved from Mhtml:file://A:\PEACE%20EDUCATION.mht!http

Galtung, J. (2003, June). Teaching and learning intercultural understanding: Fine but How? Paper presented at the UNESCO Conference on Intercultural Education, Jyvaskyla, Finland.

Kissock, C. (1981). *Curriculum planning for social studies teaching: A cross cultural approach*. Chichester: J. Wiley.

Kovalik, S. and Olsen, K. (1994). *ITI: The model integrated thematic instruction*. 3rd ed. Kent, Washington: Books for Educators, Covington Squire, ED374894.

Lasonen, J. (2004). Educating people in a culture of peace. In Lasonen, J. (Ed.), *Cultures of peace: From words to deeds. The Espoo Seminar Proceedings June 13–14, 2003* (pp. 11–13). University of Jyvaskyla: Jyvaskyla, University Printing Press.

Mickelson, R., Nkomo, M., and Smith, S. (2001). Education, ethnicity, gender and social transformation in Israel and South Africa. *Comparative Education Review*, 45 (1): 1–35.

Nagler, M.N. (1999). Peace making through violence. Retrieved from file://A:\Nagler%20Peacemaking%20through20Nonviolence.htm

Nwoji-Okeke, J.R. (1999). Teachers' strategies for effective utilization of learning resources in teaching social studies. In Nzewi, U.M. (Ed.). *The teacher: A book of readings* (pp. 269–273). Onitsha: Africana—FEP Publishers.

Ogunbi, A. (1987). Don't blame the youths. *National Concord*. April 13, 1987, p. 3.

Readon, B. (1998). *Educating for global responsibility: Teacher-designed curricula for peace education, K-12*. New York: Teachers College Press: Columbia University.

United Nations (2000). A declaration on a culture of peace. Retrieved from http://www.unesco.org/cpp/uk/projects/declarations/2000

United Nations Resolutions A/RES/51/13: Culture of Peace and A/RES/53/243, Declaration and program of action on a culture of peace.

Walker, D. (1995). *Integrative education*. [ERIC Digest–101]. Retrieved from http://chiron.valdosta.edu/whuitt/files/walker_integrated_ed.html

# Section IV
# Adult Education

CHAPTER THIRTEEN
TOWARD SUSTAINABLE PEACE EDUCATION:
THEORETICAL AND METHODOLOGICAL
FRAMEWORKS OF A PROGRAM IN SOUTH AFRICA

*Tim Houghton and Vaughn John*

**Introduction**

In this chapter, we trace the history and development of the Peace Education Program in the Centre for Adult Education (CAE) at the University of KwaZulu-Natal (UKZN), South Africa, and examine how *action research* has shaped our course. We discuss the concept of *communities of practice* and show how nurturing and supporting our trainees and students in communities of practice may contribute to sustained peace education in the province of KwaZulu-Natal.

We argue that such an examination is pertinent at this time because democracy generally, and over a decade of it in South Africa, has to be seen as a peace dividend. Given our violent history, and the lack of democracy and freedom for the majority of our people associated with that time, it is clear that democracy was possible only when a certain measure of peace was achieved. Sadly, the end of apartheid has not signaled the end of South Africa's deeply entrenched structural violence. Today South Africa remains plagued by rising poverty, unemployment, crime, and the HIV/AIDS epidemic. Despite a decade of democracy, we move steadily toward being an increasingly divided and violent society. It is in this context that efforts aimed at peace building become so crucial.

We teach a course that educates for peaceful change and that creates spaces for communities of peace practice. In this chapter, we discuss an area of peace education that is marginal in the field: an intervention aimed at community-based adult educators and development workers. While the majority of programs described in the peace education literature refers to formal schooling contexts, targeting children and youth (Hicks, 1988; Harris, 1996, 1999; Bar-Tal, 2002), our work targets adults. The marginal status of this work is confirmed by research; for example, Nevo and Brem (2002) found that between 1981 and 2000, almost 1000 articles were published on peace education. They note with concern that "Only a few PE (*sic*) programs aimed at adults" (p. 5). Despite this marginal status of peace education for adults, we wish to demonstrate through this paper that peace education can benefit from adult education's well-established tradition of "critical self-reflective practice" (Collins, 1991) and its strong links with social movements (Welton, 1995).

The essence of this chapter is an examination of the strategies and curricular approaches employed in our program that contribute to sustaining peace education in a society moving out of conflict. We suggest that action research and the notion of communities of practice as key components of a long-term peace education program may have relevance and value for similar peace education programs elsewhere.

## Context of Our Work

### A History of Political Violence in KwaZulu-Natal

The political violence which characterized sociopolitical relationships in the province of KwaZulu-Natal during the eighties, culminating in the Seven Days War of March 1990, and which subsequently so brutalized the populace well into the nineties, has been well documented (Aitchison, 2003; Butler, 1994; Greenstein, 2003; Jeffery, 1997; Levine, 1999). Taylor (2002) reports as many as 20,000 deaths by political violence in the province of KwaZulu-Natal alone between 1984 and 2002.

The devastating, long-term social toll of civil war in this province is counted in the resulting deaths, serious injuries, the interruption of employment and education, and the inevitable disruption of family and community life. Against the backdrop of this violence, the current high levels of poverty, unemployment, crime, and a serious HIV/AIDS epidemic, all contribute to a mounting climate of social decay and peacelessness.

### Current Culture of Violence in South Africa

Anyone writing about peace and peace education in South Africa must, of necessity, contextualize the narrative or study within the prevailing culture of violence (see Dovey, 1996; Maxwell, 2002). Maxwell (2002, p. 2) notes that "(d)ecades of minority rule, brutally enforced, have left a country whose people, structures and institutions bear the scars of violence, of inequality, of opportunity denied." She cites Simpson (1993), who refers to the way in which "violence in the political arena (has become) a socially sanctioned method for dealing with conflict and for achieving change." It is this legacy that manifests in the appallingly high rates of criminal and domestic violence currently experienced in South Africa. In this context, reversing such a culture of violence and educating for peaceful change is an enormous challenge.

It is this endemic violence, often with origins in a militaristic and divisive apartheid dispensation that remains to be adequately tackled by the new democratic dispensation and that confronts our Peace Education Program. A culture of violence in South Africa has grown over several decades. Likewise, growing and sustaining peace will require a long-term effort.

## Peace Education in South Africa

"Peace and friendship amongst all our people shall be secured by upholding the equal rights, opportunities and status of all" (Congress of the People, 1955). Peace as a goal is now very much part of the national rhetoric in South Africa and as such is reflected in several policy documents. See, for example, the South African Bill of Rights in the Constitution of the Republic of South Africa (Government of the Republic of South Africa, 1996) as well as the Manifesto on Values, Education and Democracy (Ministry of Education, 2002) and the National Education Policy Act (Office of the President, 1996). There is, however, still no national plan for peace education. Attempts to integrate peace education objectives have been through National Qualifications Framework (NQF)[1] critical cross-field outcomes, a few outcomes in a learning area or subject called Life Orientation (Department of Education, 2002) and a project of the national Department of Education (Ministry of Education, 2002) that addresses values in education. A small additional contribution comes through NGO interventions (Dovey, 1996; Malan, 2002), often constrained by a fragile funding environment.

Peace education programs in Africa via university-based departments are not common. The infusion of peace-related themes in general courses is, however, becoming more common (see Malan, 2002, pp. 12–13). While we support the view that peace education should target multiple audiences and contexts, the CAE has chosen to work with a historically marginalized constituency, namely, adults.

### Developing and Refining a Peace Curriculum through Action Research

The strategies and curricula approaches we now discuss evolved through an action research approach.

The action research community is not a homogenous one. There are distinct variations of approach and divides along lines of participation, ownership, and purpose (John, 2003). The combination of reflection with phases of action and planning activity within a cyclical process constitutes the defining character of action research. We employ a critical action research approach believing that research should serve purposes of emancipation and social justice. Such a critical approach is evident in the earlier work of Carr and Kemmis (cited by Cohen et al., 2000, p. 227), who saw action research as "self-reflective inquiry by participants, undertaken in order to improve understanding of their practices in context with a view to maximizing social justice." This critical theory orientation to research relates well to a transformative pedagogy (discussed later in the chapter) aimed at "furthering human emancipation and overcoming domination and repression" (Collins, 1991, p. 104).

Cohen et al. (2001, p. 227) note that "action research is designed to bridge the gap between research and practice, thereby striving to overcome the perceived persistent failure of research to impact on, or improve, practice." We now show how action research has served the conceptualization and development of our peace education program.

## Reflections on Curriculum

The first Peace Education course offered by the Centre for Adult Education in 1998, *Towards Creating Peaceful Communities*, was a direct response to the context of political violence and intolerance that still prevailed in the province at the time.

An independent evaluation of this first course (Verbeek, 1998), then became an important moment of reflection in the action research cycle as it made a number of useful recommendations that the current course attempts to honor.

KwaZulu-Natal still experiences outbreaks of politically motivated violence. Election time and periods of floor crossing in parliament are potentially volatile moments for communities with a history of political intolerance. However, the context of conflict in the KwaZulu-Natal Midlands has generally shifted since the eighties and nineties when it was predominantly political and physically violent, to a context of competition for scarce resources around development issues. This competition inevitably creates conflict (which can also erupt in physical violence) which hampers the development of disadvantaged communities. Two key assumptions that inform our current work with communities is that if development is to be sustainable, it needs to be participatory, and that it needs to take place in a peaceful context.

## Two Models of Delivery

Our new peace education course reflects two, closely related models of delivery (see figure 13.1). The course is offered formally as a credit-bearing specialization module within a two-year, part-time, higher certificate in education. Students attend one- and two-day contact sessions over a full semester, at the university. These students proceed to Development in Practice (DIP), a service-learning module, in the final semester of their study.

The course is also offered nonformally as two five-day residential workshops supplemented by a year-long follow-up and support program, in which CAE staff maintain regular telephonic and face-to-face contact with trainees as they implement peace education/peace building projects in their own communities. Participants in this model have the option of gaining formal accreditation.

Reflecting on the pilot year of the current course, program staff grappled with what we perceived as a deficiency in the nonformal model that is difficult to resolve. The formal model clearly provides an advantage for learners because of a foundational year of study in adult education and participatory development. These learners also enjoy the benefit, during the service-learning module, of authentic development practice in a host organization. This would appear to put the nonformal participants of the course at a disadvantage. However, our close attention to learner feedback and the literature on adult learning has helped to shape the intensive year-long follow-up that provides a community of practice (discussed more fully in the chapter part that follows).

It is this kind of support that contributes, we believe, to redressing this apparent disadvantage in the form of regular site visits, focus groups, guided practice, reflection and sharing sessions, and short refresher workshops that are key components of this model.

| Peace Education Program 2006 ||
|---|---|
| *Two models of delivery* ||
| **Formal**[2] | **Nonformal** |
| As part of a Higher Certificate in Education (Participatory Development) | As part of University-Community Development Initiatives |
| **Assessment**: Compulsory | **Assessment**: Optional |
| **Phase 1** ||
| Experiential, participatory workshops ||
| 1- and 2-day workshops on campus | 5-day residential workshop |
| **Workshop 1** ||
| <ul><li>Peace and sustainable livelihoods</li><li>Peace and personal development</li><li>Alternatives to Violence Project (AVP)</li><li>Analyzing conflict</li></ul> ||
| **Project**: Analyzing Conflict ||
| **Workshop 2** ||
| <ul><li>Foundational skills</li><li>Negotiation</li><li>Mediation</li><li>The Zwelethemba Model of peace making and peace building</li><li>Educating for peaceful change</li></ul> ||
| **Project**: Educating for Peaceful Change ||
| **Phase 2** | **Phase 2** |
| Development in practice: Service-learning module | One year of follow-up support/mentorship |

**Figure 13.1** CAE Peace Education Program

## Strategies and Curricular Approaches for Sustaining Peace Education

Having discussed the conceptualization and development of the current course we now focus on its key strategies and curricular approaches.

### Content Focus

The content focus of the course has obviously shifted since 1998 with the different context of violence but the emphasis on nonviolent responses to conflict remains an important element. The course content (see figure 13.1), in many ways reflects what could be expected of a variety of peace education programs in a diverse number of settings (Burns and Aspeslagh, 1983; Harris, 1999, 2002; Evans et al., 1999). Gender relations and the environment, holistic peace, respect for diversity, foundational skills in the creative transformation of conflict, and issues of governance and citizenship are core frames of reference in our course.

The concern that peace education students "should be critically aware of the realities of structural violence and how a philosophy of development based on justice is a preferred alternative" (Castro, 1999, p. 167), is common in the literature (Toh Swee-Hin, 1988; Summy, 1995; Harris, 1999, 2002). Our emphasis on participatory development is central and gives coherence to the whole course. The course also aims to equip participants with the skills to facilitate learning and is specific to the context of community educators and development workers.

### Experiential Learning

Experiential learning is commonly cited as a key method of peace education delivery (Bar-Tal, 2002, p. 8) and is well established within broader adult education practice (Kolb, 1984; Bateson, 1994; Boud, Keough, and Walker, 1996; Usher, Bryant, and Johnson, 1996).

In our educational practice, experiential learning comes with its own set of contextual and pedagogical predicaments. Providing learners with an experience and trying to get them to reflect on that, does not automatically lead to critical thinking and attitude and/or behavior change. Educators, who assume that a learning experience in the classroom, however powerful and transformative, will produce praxis, are likely to be disappointed. In fact, experiential learning can be as domesticating as traditional "chalk and talk" methods of teaching, unless it is problematized (Grundy, 2001; Freire, 1972). Facilitation of such learning methods requires a very experienced facilitator who is well prepared and, in our context, ideally able to communicate in both English and the vernacular language for the most effective communication and learning to take place (Hlela, personal communication). This is seldom the reality.

We attempt to draw on and make explicit the collective dynamic of learning that is integral to the African experience—that we learn with, from and through others, and that the ability to utilize the experience of other people and awareness of how people pool expertise and rely on each other should be understood as central to the way in which experiential learning happens.

## Transformative Learning

Experiential learning components specifically designed to facilitate transformational learning are important in our pedagogy.

Transformative learning is a constructivist theory of adult learning that postulates that the way people interpret their experience depends on understanding learnt through early socialization (Mezirow, 1991). These socializations become frames of reference through which we understand our experiences. People generally have a strong tendency to interpret experiences in ways that fit their established understanding, and to avoid having this understanding challenged. "When circumstances permit, transformative learners move toward a frame of reference that is more inclusive, discriminating, self-reflective, and integrative of experience" (Mezirow, 1997, p. 5).

Mezirow (1991, 1997) has been criticized by Pietrykowski (1998) for an essentialist approach to learning that has perspective transformation happening at the "micro-level of personal transformation rather than at the macro-level of social and political change" (p. 69). While our internal evaluations suggest that participants of our course experience significant personal transformation, which is undeniably pleasing, we understand that such personal transformation does not necessarily translate into social transformation. Indeed it would be naïve to believe that a ten-day workshop, despite the emphasis on educator training, would produce practitioners capable of activating and sustaining significant social change in their communities.

However, prolonged support and encouragement for trainees subsequent to their training, in actual peace education and peace building initiatives, may provide a community of practice that could foster a translation of personal transformation into the broader social transformation necessary for sustaining peace education. We explore this more fully in the chapter parts that follow.

## Partnerships

To ensure the long-term sustainability of the Program, a range of partnerships have been considered, including partnerships with local government and the corporate sector. In addition, the current course has partnerships/alliances with the Alternatives to Violence Project (AVP), the African Centre for the Constructive Resolution of Disputes (ACCORD) and the Community Peace Program (CPP).

### *Alternatives to Violence Project (AVP)*

Workshop One of our course includes a complete two-day AVP Basic Workshop that carries its own certification, career pathway, and a national and local support network. The Alternatives to Violence Project (AVP) is a grassroots, international, volunteer movement committed to reducing interpersonal difficulties in society and based on the peace-building principles of among others, Gandhi and Martin Luther King. AVP works toward its goals by presenting experiential workshops in business, government, prisons, schools, communities, and other organizations.

### *Community Peace Program*

Workshop Two of our course exposes/introduces participants to the Zwelethemba model of peace making and peace building. This model has been developed in

South Africa as a tool for grassroots organizations (formally constituted Peace Committees) to assume responsibility for local democratic governance. The Community Peace Program (CPP) also offers our course participants a career pathway and a national and local support network.

Although our university does not have a strong history and tradition of collaboration with broader civil society, cooperation between the CAE and other organizations outside the academy, such as AVP and CPP, has worked well in drawing NGO expertise into our program.

The alliances with other NGO providers of peace education ensure that our course is fresh and draws on the experience and knowledge of practitioners in communities. Wilson (2001) has encouraged the forging of professional alliances as a strategy for building the professional identity of adult education.

## Communities of Practice

### The Need for Authentic Practice

Clearly, action research has been pivotal to curriculum development and change within our Peace Education Program. As explained earlier, it has helped shape the current course in the most significant ways. More recently, our reflection on evaluation feedback and the nature of support being requested by learners has steered us into recognizing that what our learners appear to need and request are spaces for learning in and through authentic practical activity.

In 2003, we piloted a postcourse support program with a group of learners. During the postcourse support phase, learners were expected to undertake modest peace education activities within their communities. Interestingly, the peace-related activities that these learners most commonly were able to take forward or implement within their communities were aspects of the course such as AVP and CPP (discussed earlier). Both these aspects of the course have a set of strikingly similar features, namely, they have well-developed and defined curricula or implementation processes; they come with well-established support networks of practitioners and they enjoy established identities as programs that exist in other parts of South Africa or the world.

We had hoped that many of the educators in this group would have implemented with their own learners other aspects of the peace education program to which they had been introduced. A year later, we found that, despite offers of support and visits to their communities, they had, only to a limited degree, been able to practice peace education. We know that many of the learners trained in 1998 had also not been practicing as peace educators as we had hoped. Other NGOs offering peace-related training, for example, the Independent Projects Trust, have reported a similar lack of training returns (Goodenough, 2004). During an evaluation session with a group of learners who had completed the formal course in 2004, the establishment of a "peace club" was suggested as a means of supporting learners. We believe that the value of a number of training programs is compromised through the lack of opportunity for trainees to use their knowledge and skills in authentic and meaningful ways within a supportive environment. With many of these courses, the providers and their curricula focus mainly on what happens during the course. Little attention is paid to the postcourse context into which participants return. In the past, our own course had suffered from a similar narrow conception of peace curriculum.

Deeper reflection and engagement with learning theory literature has allowed us to see possible connections between various examples of learner feedback as presented above and with the theory of communities of practice. In this part of the chapter, we therefore explore how communities of (peace) practice may serve the goals of sustained peace education and educator/peace worker development.

### Challenges in the Field of Adult Education

There is sufficient evidence that learning that takes place on a course is best retained when it is followed by practice and mentoring (Joyce and Showers, 1988). For peace education in schools, educators have a readily available site for practice and mentoring opportunities. In the adult education context in South Africa, such space for practice and mentoring is rare. The problem facing our novice educators is, "Where do they acquire authentic, hands-on practice of peace education?" A further problem is that there are no institutional fora supporting the practice of peace education. Professional associations for adult educators in South Africa have disappeared. There are no places for community-based peace educators such as those that we work with, to come together to talk about their practice, support each other and grow their confidence, knowledge, skills, and commitments. We believe that these factors seriously limit long-term peace education efforts.

Therefore, in the South African context, where adult education provision is scarce and peace education is not a common or mandatory part of the formal and nonformal curricula, there are few opportunities for new peace educators to practice and hone their skills and to develop the identity of "peace educator." The context is unlike that where peace education is a standard part of school curriculum and where educators can find sites of practice and groupings and organizations that support identity development, for example, professional teacher associations, peace educator networks and conferences. The peace education community in South Africa is simply too small and too dispersed to provide an academy for its new recruits. How then does one grow this field? We believe that one option is to facilitate access to a "community of practice" within our peace education program for initiating, developing, and nurturing new peace educators. We believe that in different ways through their feedback students have been indicating this need.

### The Communities of Practice Theory

The notion of communities of practice arises from the work of Jean Lave and Etienne Wenger (1999) and Wenger (1998). According to Lave and Wenger, learning is the increasing participation in a "community of practice." The process of learning begins with "legitimate peripheral participation" where a newcomer is allowed access to a practice but spends some time at the periphery of the practice and then gradually moves to the center of the practice and becomes a full participant. The newcomers (or novice peace educators in our case) eventually make the practice their own and thus become old-timers. The use of artifacts with understanding of their significance to the practice and the assimilation of the identity of the practice are two central parts to this process of learning through and in activity. Lave and Wenger (1999, p. 22) state that

> To begin with, newcomers' legitimate peripherality provides them with more than an "observational" lookout post: It crucially involves participation as a way of learning—of both absorbing and being absorbed in—the "culture of practice." An extended period of legitimate peripherality provides learners with opportunities to make the culture of practice theirs.

This model of learning proposes that learning is a social process that takes place through a process of participating in a community of practice. According to Lave and Wenger (1999, p. 22) "legitimate peripheral participation" provides a way to speak about the relations between newcomers and old-timers, and about activities, identities, and communities of knowledge and practice. We believe that this idea of learning in and through activity has value and relevance to peace education and the development of new peace educators.

## Attempts to Develop and Nurture a Community of Practice

The year of postcourse support in our Peace Education Program could be a central opportunity for introducing and engaging the learner within a community of practice. Through this opportunity to learn through peace activity within a community the learners can practice and implement some of their new knowledge, values, and skills. They can do this within a socially supportive environment. In the process, they learn the practice of the community and gradually make such practice their own. They are thus immersed within a culture of peace and can adopt such a culture. With the next group of learners, we are hoping to strengthen the sense of a community of practice by offering learners a peace education "starter pack." The learners would be supported in implementing it in their respective communities. We hope to create spaces where they can come together after and during such implementation to share their experiences and to move closer to the center of the practice. Lave and Wenger (1999, p. 3) state, "The fact that they are organizing around some particular area of knowledge and activity gives members a sense of joint enterprise and identity." In writing about the role of adult educators, Collins (1991) has likewise argued that "pedagogical strategies should be adopted that further collective decision-making and solidarity in students" (p. 109).

The model of learnerships currently being used by Sector Education and Training Authorities (SETAs) in South Africa[3] has some similarity to what is being proposed with the notion of communities of practice. However, the SETA model tends to be strongly associated with professionalization of fields of practice, which, over the last few decades, has caused considerable concern amongst some adult educators (Cruikshank, 1993). We prefer a model of community of practice that is less institutionalized and more closely aligned to a social movement ethos. In some contexts, relevant social movements such as peace movements, human rights movements, and women's movements may also provide opportunities for learning in communities of practice. In calling for an enlarged view of learning, Overwien (2000) has pointed to the opportunities created by social movements for informal learning of important skills and competences. Such learning allows for closer links between learning and acting or practice.

We recognize that both our ideas about communities of peace practice and existing learnership models do not sit comfortably within a university context. University rhetoric reflected in vision and mission statements often professes commitment to excellence in programs of development by mounting appropriate curricula. However, university courses and administrative systems are still designed for traditional students who are usually young, recent school-leavers who study full-time through attending lectures on campus. Programs such as ours, which include learning through communities of practice, pose a serious challenge to this model in that our learners are mature, part-time learners who also learn through situated activity in communities and organizations outside of the university campus. University funding models also do not cater to such nontraditional provision. It is time for universities to seriously rethink their models of provision and to cater to a wider range of students and learning environments.

## Conclusion

With the end of apartheid and a little more than a decade of democracy, South Africa remains a deeply divided and violent country. In a context where peace education opportunities for adults are rare, the quest for sustained peace education becomes paramount. Short-term peace education interventions have limited impact. Peace Education, which intends to move beyond knowledge and skills acquisition to including peace action, requires more sustained peace education programs and growth of the peace educator community in South Africa.

This chapter argues that the field of adult education has a rich base of theoretical and methodological tools for peace education to draw on. We have focused on two of these tools, namely, *critical action research* and *communities of practice* and have examined how these tools continue to shape a peace education curriculum. We believe that a transformative pedagogy that taps the benefits of action research and communities of practice can yield a curriculum for sustained peace education. Where social transformation is a goal, critical action research is a powerful methodology for keeping curricula relevant to context and learner needs. Communities of practice can provide a robust theoretical space for understanding and developing support structures for novice peace educators.

## Notes

1. The South African Qualifications Authority uses the National Qualifications Framework (NQF) to approve, register and monitor the quality of all education and training in South Africa.
2. Prior to this, students would have completed one year of foundational study in adult education and development.
3. South Africa has established 25 Sector Education and Training Authorities (SETAs) that regulate education and training for different industries and sectors.

## References

Aitchison, J.J.W. (2003). The origins of the Midlands war. In Greenstein, R. (Ed.), *The role of political violence in South Africa's democratisation.* (pp. 47–72). Johannesburg: Community Agency for Social Enquiry.
Bar-Tal, D. (2002). The elusive nature of peace education. In Salomon, G. and Nevo, B. (Eds.), *Peace Education: The concept, principals and practices around the world* (pp. 27–36). Mahwah, N.J.: Larry Erlbaum Associates.
Bateson, M.C. (1994). *Peripheral visions: Learning along the way.* New York: HarperCollins.
Boud, D., Keough, R. and Walker, D. (1996). Promoting reflection in learning: A model. In Edwards, R., Hanson, A. and Raggat, P. (Eds.), *Boundaries of adult learning.* New York: Routledge.
Burns, R. and Aspeslagh, R. (1983). Concepts of peace education: A view of western experience. *International Review of Education*, 29: 311–330.
Butler, M. (1994). *Natal, violence and the elections.* Pietermaritzburg: Centre for Adult Education, University of Natal.
Castro, L.N. (1999). Peace and peace education: A holistic view. In Perez de Cuellar, J. and Young, S.C. (Eds.), *World encyclopedia of peace*, 2nd ed. (Vol. IV, pp. 164–171). New York: Oceania Publications.
Cohen, L., Manion, L. and Morrison, K. (2000). *Research methods in education*, 5th ed. London: Falmer.
Collins, M. (1991). *Adult education as vocation: A critical role for the adult educator.* London: Routledge.
Congress of the People (1955). The freedom charter. Proceedings of the Congress of the People, Kliptown, South Africa, June 26, 1955. Retrieved October 17, 2005 from http://www.anc.org.za/ancdocs/history/charter.html
Cruikshank, J. (1993). The role of advocacy: A critical issue in Canadian university extension work. *Studies in the Education of Adults*, 25: 172–184.
Department of Education (2002). *Revised curriculum statement. Grades R–9 (schools) policy.* Life Orientation. Pretoria: Department of Education.
Dovey, V. (1996). Exploring peace education in South African settings. *Peabody Journal of Education*, 71(3): 128–150.
Evans, D. et al. (1999). *A culture of peace. Educational innovation and information.* 100. Geneva: UNESCO International Bureau of Education.
Freire, P. (1972). *Pedagogy of the oppressed.* Harmondsworth: Penguin.
Goodenough, C. (2004). *Post conflict peace building in Richmond.* Durban: Independent Projects Trust.
Greenstein, R. (2003). *The role of political violence in South Africa's democratisation.* Johannesburg: Community Agency for Social Enquiry.
Government of the Republic of South Africa. (1996). Constitution of the Republic of South Africa: Bill of Rights. Retrieved October 11, 2005 from http://www.gov.za/constitution/1996/96cons.htm
Grundy, S. (2001). *Curriculum: Product or praxis?* London: Falmer Press.
Harris, I. (1996). Peace education in an urban school district. *Peabody Journal of Education*, 71(3): 63–83.
Harris, I.M. (1999). Types of peace education. In Raviv, A., Oppenheimer, L. and Bar-Tal, D. (Eds.), *How children understand war and peace* (pp. 299–310). San Fransisco: Jossey-Bass.
Harris I.M. (2002). Conceptual underpinnings of peace education. In Salomon, G. and Nevo, B. (Eds.), *Peace education: The concept, principals and practices around the world* (pp. 15–26). Mahwah, N.J.: Larry Erlbaum Associates.
Hicks, D. (Ed.), (1998). *Education for peace: Issues, principles and practices in the classroom.* London: Routledge.
Jeffery, A. (1997). *The natal story: 16 years of conflict.* Johannesburg: South African Institute of Race Relations.

John, V.M. (2003, October–November). Educational "action research" partnerships: Whose action and whose research? Paper presented at the Kenton Conference, Worcester, Western Cape.
Joyce, B. and Showers, B. (1988). *Student achievement through staff development.* London: Longman.
Kolb, D.A. (1984). *Experiential learning: Experience as the source of learning and development.* Engelwood Cliffs, NJ: Prentice Hall.
Lave, J. and Wenger, E. (1999). Legitimate peripheral participation in communities of practice. In McCormick, R. and Paechter, C. (Eds.), *Learning and knowledge* (pp. 21–35). London: Paul Chapman.
Levine, L. (1999). *Faith in turmoil: The seven days war.* Pietermaritzburg: Pietermaritzburg Agency for Christian Social Awareness.
Malan, J. (2002, October). Education for peace in Africa: Review of activities in the field of education for peace already in progress in Africa. Paper presented at the Advisory Meeting on the Africa Programme of the University for Peace (UPEACE), Maputo, Mozambique.
Maxwell, A. (2002, July). Educating for peace in the midst of violence: A South African experience. Unpublished paper presented at IPRA conference, Seoul, South Korea.
Mezirow, J. (1991). *Transformative dimensions of Adult learning.* San Francisco: Jossey Bass.
Mezirow, J. (1997). Transformative learning: Theory to practice. New *Directions for Adult and Continuing Education,* 74, Summer 1997: 5–12.
Ministry of Education (2002). *Manifesto on values, education and democracy.* Cape Town: Cape Argus Teach Fund for Department of Education.
Nevo, B. & Brem, I. (2002). Peace education programs and the evaluation of their effectiveness. In Salomon, G. and Nevo, B. (Eds.), *Peace education: The concept, principals and practices around the world* (pp. 15–26). Mahwah, NJ: Larry Erlbaum Associates.
Office of the President. (1996). National Education Policy Act. No. 27 of 1996. South African Qualifications Authority. Retrieved October 11, 2005 from http://www.saqa.org.za/show.asp?main=docs/legislation/related/act27-96.htm&menu=home.
Overwien, B. (2000). Informal learning and the role of social movements. *International Review of Education,* 46: 621–640.
Pietrykowski, B. (1998). Modern and postmodern tensions in adult education theory: A response to Jack Mezirow. *Adult Education Quarterly,* 49: 67–69.
Simpson, G. (1993). Women and children in violent South African townships. In Motshekga, M. and Delport, E. (Eds.), *Women and children's rights in a violent South Africa* (pp. 3–13). Pretoria West: Institute for Public Interest, Law and Research.
Summy, R. (1995). Vision of a non-violent society. In M. Salla et al. (Eds.), *Essays on peace* (pp. 63–69). Rockhampton: Central Queensland University.
Taylor, R. (2002). *Justice denied: Political violence in KwaZulu-Natal after 1994.* Johannesburg: Centre for the Study of Violence and Reconciliation.
Toh Swee-Hin (1988). Justice and development. In Hicks, D. (Ed.), *Education for peace: Issues, principles, and practice in the classroom* (pp. 122–142). New York, London: Routledge.
Usher, R., Bryant, I. and Johnson, R. (1996). *Adult education and the postmodern challenge: Learning beyond the limits.* New York: Routledge.
Verbeek, C. (1998). *Report on an independent evaluation of: Towards creating peaceful communities: A peace education course.* Commissioned by the Centre for Adult Education, University of Natal, Pietermaritzburg.
Welton, M. (1995). Amateurs out to change the world: A retrospective on community development. *Convergence,* 28: 49–53.
Wenger, E. (1998). *Communities of practice. Learning, meaning and identity.* New York: Cambridge University Press.
Wilson, A.L. (2001). Professionalization: A politics of identity. *New Directions for Adult and Continuing Education* (Fall): 73–83.

# Chapter Fourteen
# Learning and Unlearning on the Road to Peace: Adult Education and Community Relations in Northern Ireland

*Paul Nolan*

In November 2002, the United Nations Special Rapporteur on the Right to Education, Katarina Tomasevski, paid a visit to Northern Ireland to examine the role of education in rebuilding a society that had been torn apart by a conflict that had lasted for almost 30 years and claimed the lives of three and a half thousand people. She saw that the signs of progress real and visible. Across the skyline of Belfast she could see the cranes and scaffolding of new building programs, visual metaphors for the wider, societal changes then underway. Noting that the European Union special funding programs driving the reconstruction was of a temporary nature, the main recommendation of Tomasevski's report was clear. The EU Peace Program ". . . may have temporarily increased employment, but education constitutes the long-term solution" (Tomasevski, 2003, p. 7).

Faith in education as a solution to ethnic conflict can almost be taken as a given, particularly amongst educationalists. Our a priori assumptions however are not always matched by an equal confidence that we know HOW education assists reconciliation processes, or the ways in which the efficacy of different approaches can be contrasted and compared. In addition we have to admit the heretical thought that when we speak of education not as an abstract, but as a concrete set of social institutions and embedded cultural practices, then it is possible that in particular conflict situations education is part of the problem, not the solution. The British Government Department for International Development, in a 2003 report, *Education, Conflict and International Development* (Smith and Vaux, 2003) drew both upon academic literature (Bush and Saltarreli, 2000) and observations from various theaters of war to draw attention to the "two faces of education." The negatives on the balance sheet include the use of education as a weapon in the cultural repression and exclusion of minority populations (as in Guatemala); segregated schooling systems that perpetuate divisions (as in South Africa); the manipulation of history in textbooks to create the "national story" (as in the former Soviet republics); and the closing off of access to educational facilities (as in Israel's forced closure of educational institutions in the Palestinian territories).

Northern Ireland could serve as a further case in point, where the benign effects of education are often offset against the unintended consequences of structural segregation. The purpose of this chapter will be to look at how those working in adult education cope with the Sisyphean struggle of peace building in a society where the deep divisions are created in part by the education system itself.

## Education and Social Structure in Northern Ireland

Northern Ireland is, famously, not one community but two. Even the peace accord known as the Good Friday Agreement starts off by talking about "the community" but slips quickly into language describing "the two main communities." The origins of this historic divide have been well rehearsed elsewhere by various authoritative texts (see, for example, Bardon, 2001 and Foster, 1990) dealing with what has been described as "this most over-researched of conflicts" (Pollak 1993). In essence, the Plantation of Ulster in the seventeenth century left a settler group of Protestants in an uneasy relationship with the indigenous Irish Catholic population, and the partition of Ireland in 1921 simply reversed the majority/minority ratios in the North East, leaving the Protestants of Ulster as the dominant group within the new six-county state of Northern Ireland, while remaining the minority on the island as a whole. Protestants tend to describe themselves as unionists because they wish to maintain their link with the British state; Catholics on the other hand tend to identify with the nationalist aspiration of a united Ireland, freed from any constitutional links with Britain. The first proposals for schooling brought forward by the newly created Northern Ireland Parliament in 1923 were for a religiously neutral form of state provision, but secularism was not on the agenda of either of the two main blocs, and a rapprochement that allowed state schools to teach religion and become de facto Protestant was reached. Catholic schools, however, maintained their independence with, until recently, lower levels of state support. The first integrated school, Lagan College, was opened by parents in 1981 with 21 pupils, with the overt intention of creating an alternative to the binary divide. Since the Education Reform (NI) Order of 1989, it has been official government policy to encourage integrated education. There are now around 50 schools in total, 18 of which are second level colleges and 32 are integrated primaries. The achievement is admirable, but to provide some perspective, after a quarter of a century, the total only accounts for 5 percent of the school-going population. The other 95 percent is still in schools that are identified, and identify their pupils, as being either Catholic or Protestant.

This degree of segregation would not in itself be remarkable—after all, denominational schools exist side by side throughout the world—but for the fact that it maps directly on to the deeper historical fault line. Of crucial importance is the degree of residential segregation. The map of Northern Ireland is constructed out of a patchwork of ethnonationalist identities where, at some interface zones, physical walls separate Protestant and Catholic neighbors. There are over 30 such "peace walls" in total, with the majority of them in Belfast, where over 60 percent of the population lives in areas that are internally homogenized to the extent that 90 percent of their neighbors are of the same religion. That same mutual mistrust also plays itself out in patterns of production and consumption. A report from the Equality Commission (2001) showed that there were continued and high levels of segregation across all private sector businesses. A year later, a report on "chill factors" (Shirlow et al., 2001) showed that half of those sampled had experienced intimidation in the workplace due to their religious affiliation, and that fear was a major factor for people in determining where they would seek employment.

The dates are important here: paradoxical as it might seem, the Good Friday Agreement of 1998 would appear to have accelerated a movement of the two communities *away* from each other, rather than–as was hoped at the time–bringing forth new forms of reconciliatory politics. While the Agreement put an effective end to the politics of armed struggle, the tectonic plates were moved further apart by the primacy given to ethnic, rather than civic identity. This quickly became apparent, for example, in the turf wars over parades. The centrality of education to the conflict was brought home by the sight of angry mobs of adults threatening small children as they attempted to walk to the Holy Cross School in Ardoyne in September 2001. Voting patterns since 1998 confirm this sense of polarization. In 1997, the Reverend Ian Paisley's Democratic Unionist party accounted for 15.8 percent of the vote in the Westminster elections, and by May 2005 it had increased its share to 33.7 percent. At the other end of the spectrum, Sinn Fein moved to become the party of choice for most nationalists, increasing its share from 16.95 percent to 24.4 percent. The good news was that the conflict had been transformed from being a bloody, low-level civil war to something that could be fought out at the political and cultural level; the bad news was that the centre ground had collapsed and the poles had been magnetized, splintering the body politic into two distinct monocultures.

The obvious problem with such monocultures is their tendency to act as hothouse cultures for the growth of more virulent strains of sectarianism. It comes almost as early as conscious thought, and its basic patterning can be detected even before the preschool stage. Studies done with children as young as three show them already building a mental framework around "us and them" identities. Positive values are ascribed to communal symbols (flags, football shirts, etc.) with a corresponding negativity attaching to the symbols of the other side (Connolly, 2003). With a force field shaped by such strong and mutually reinforcing determinants, education has to fight to find neutral ground. Segregation in schooling is both cause and consequence, each acting upon the other within a sectarian loop. The emancipatory mission of education therefore has as its first obstacle the schooling system itself. Thus, for example, as the Northern Ireland Council for Integrated Education points out, the difficulty that attends those initiatives that attempt to bring Protestant and Catholic children together is that it is inherently self-defeating, as the children in segregated schools enter the process bearing different identities:

> Operating in this context . . . runs the risk of reinforcing the divisions that exist in Northern Ireland by setting up the "other" community as an entity which exists "out there" and which is to be engaged with (directly or indirectly) only in carefully structured and planned activities after which everyone can settle back into the security of their own isolated identity. (NICIE)

In a quite separate context, but one which has relevance here, Gutman (1995) points out that tolerance is too weak an ideal to power forms of civic education in the United States. If white and black people—or, in this context, Catholics and Protestants—tolerate but do not interact with each other, then the most benign outcome is a "live and let live" scenario where divided communities tolerate each other's presence, but do not employ each other or socialize together, and no attempt is made to address structural inequalities. A more radical approach is required, or perhaps it is more accurate to say that the complex of problems that are tangled together in the Northern Ireland situation require many different approaches at many different levels. It is also true to say that over the past 30 years Northern Ireland has functioned as sort of laboratory that has allowed various educational theories to be tested—and tested in the most demanding of circumstances.

**Adult Education and Peace Building**

It has been the shared ambition of the adult education community in Northern Ireland to contribute to peace building (Department of Education for Northern Ireland, 1998; Department for Employment and Learning, 2004). To greater and lesser degrees that ambition is to be found in its 16 colleges of further education, its 3 universities, and is perhaps most visibly present amongst those voluntary adult learning providers such as the Workers' Educational Association and the Ulster People's College which have taken community relations right to the core of their mission. Beyond these formal providers lie a myriad community-based organizations negotiating and constructing curricula to address very local interests. What type of courses result? A glance through course programs and publicity brochures for community education will regularly display the following activities: prejudice reduction courses, local history classes, assertiveness training, victim support programs, equality awareness, mediation training skills, antisectarian workshops, listening skills training, capacity building for voluntary organizations, classes in Ulster Scots and in the Irish language (sometimes offered to both communities), tin whistle or pipe band training, workshops in managing grief and loss, community development training, conferences on Irish history and countless more. If we are to go beyond those courses with an overt claim on addressing issues of division to include, as the EU does, those programs that rebuild communities through vocational training or initiatives to combat social exclusion, then we quickly lose the means to draw a circumference around this concept. How do we set the boundaries and how do we begin to map the field to understand better what is going on?

One starting point might come from a distinction drawn by Kember et al. (2001), which posits the view that concepts of adult education teaching remain polarized between "teaching as transmission of knowledge" and "teaching as learning facilitation." If we are to attempt a diachronic as well as a synchronic analysis of community relations education in Northern Ireland, the narrative arc would describe a movement from the first model, that of knowledge transmission that was dominant at the start of the Troubles toward the more exploratory and experiential forms of learning that developed at grassroots community level.

When the Troubles first appeared on television news bulletins across the world in the late 1960s, the curiosity was that of a sixteenth-century religious war erupting onto what appeared like modern housing estates, and the "myth of atavism" (Vincent, 1993, p. 123) provided the explanatory text for what was happening. The role of education appeared obvious in a situation where Enlightenment assumptions provided not only an explanation of the problem, but a practical solution to be applied: knowledge could be harnessed in the fight against prejudice, and the premodern attachment to myth, religion, and tribal identity could be dissolved in the healing bath of reason. The faith that "Knowledge is Power" takes us back, of course, to the taproot for adult education historically—it was in fact the slogan of the early Workers' Educational Association (Jennings, 1979)—and in the early days of the conflict, it was thought useful to offer courses that could "explain" the roots of the problem, in the hope that a greater understanding of Irish history and society might in itself act as a counterweight to prejudice.

Such an approach appears innocent now in the disarray of the postmodern world. The rise of ethnic tensions worldwide and, in particular, the televised collapse of the former Yugoslavia makes the Northern Ireland conflict appear, in retrospect, not so much an historical throwback as a harbinger of things to come. As the writer and critic Edna Longley puts it, "Thirty years ago I used to wish that Northern Ireland would get more like the rest of the world; today I fear that the rest of the world is getting more like Northern Ireland" (Longley, 1993, p.8). The new politics of identity have reconfigured the context for debate about ethnonationalist cultures, and, on the way, the belief in the power of objective knowledge has been abandoned. If further proof were required, the deepening of sectarian identity that has gone alongside the massification of higher education would seem to indicate that access to other systems of knowledge does little to soften attitudes. Over 45 percent of school leavers in Northern Ireland go on to study at university; the new leadership cadres in the main political parties are all graduates, and yet here, as indeed in Serbia, Bosnia and Croatia and in virtually every other trouble spot on the globe, knowledge does not lead to the sort of enlightenment and tolerance imagined by the founders of the adult education movement a century ago. The ethnic entrepreneurs who mobilize popular discontents tend to have after their names not just letters but undimmed prejudices too.

It would be easy therefore to join what Brian Barry (Barry, 2001) calls the "flight from Enlightenment," and to declare—as so many have done—that the "Enlightenment project" has failed (Macintyre, 1985; Gray, 1995). Educationalists do not have this option. Belief in the quest for knowledge and truth is intrinsic to the mission. What has to be allowed however is that rationality will not, of itself, deliver on social goals. Something else is required. If attitudinal change is the aim, then affective learning must take place and, in such a highly segregated society, E.M. Forster's famous injunction—"Only Connect"— provides its own justification. There have been countless cross-community efforts to bring people together at local level, sometimes to counter negative stereotypes indirectly though sporting or cultural activities, sometimes to allow the problem to be faced head-on in discussion groups, facilitated by trained tutors or mediators. Belief in "contact theory" (Allport, 1954; Pettigrew, 1998) is such that a number of residential centers have been created to facilitate such cross-community exchanges: the Corrymeela Centre on the north coast is perhaps the best known, and has facilitated both youth and adult exchanges since 1965 when it was founded to promote a new sense of ecumenism that was briefly in the air. In addition, the churches, trade unions, and voluntary organizations regularly use the residential experience to facilitate exchanges between Protestants and Catholics. The reconciliation of different world views is the assumed (but seldom stated) intention of such activity, and the larger political issues are subordinated in order to retain the focus on improving person-to-person or community-to-community relationships. The commonsense nature of this approach does not shield it from criticism. It is, according to Ryan, "the crudest version of peace-building" (Ryan, 1995, pp. 223–265) and Cochrane and Dunn (2002, p. 78) explain the failings of the contact hypothesis in these terms: "Such analysis stresses dysfunctional human relationships as causal factors of sectarianism and conflict, rather than historical forces or malign political structures."

Evidence from research tends to support the skeptical view. While individual relationships can and do develop between Protestants and Catholics who meet in the contrived atmosphere of the contact group, no discernible change takes place at-community-to-community level. Trew (1986, pp.104–105) summarized the evaluations of various contact programs by saying that "there is no empirical evidence or theoretical rationale to suggest that contact *per se* will either influence salient political beliefs or have any impact on sectarianism in the society." That cold assessment has been endorsed by later studies, as in the summary of the literature by Hargie and Dickson (2003, p. 295) that concludes that on occasion the results of contact work can be negative as well as positive: the more we meet others the *less* we may like them. Contact, they suggest, may be necessary but it is not sufficient.

What other strategies then have proved viable? Various typologies can be created, and within them various categories and subcategories, but I am happy to accept the simple distinction proposed by Ifat Maoz in his study of 47 different peace education programs undertaken in the Middle East between 1999 and 2000. Maoz (2002, pp. 259–271) divides the initiatives into two basic types: those that aim to increase understanding and foster coexistence (the knowledge-based and contact-based approaches described above would fall into this category) and, at the other end of the spectrum, more confrontational models designed to expose inequalities and empower those who experience injustice. This model can easily transfer to the Northern Ireland experience with some adjustments in terminology: instead of using the term confrontational, the label of antisectarian is used to describe those education and training initiatives that aim to challenge the sectarian biases within institutional structures and practices, and to examine the inequities of the existing power structures. The most significant developments have occurred within workplace settings and have been assisted by a very strong legislative framework. Under the terms of the Northern Ireland Act (1998), not only are all forms of discrimination outlawed, but public sector employers have a responsibility to actively promote good relations between the two communities.

The trade unions have often been central to educational initiatives that attempt to examine the forms of symbolic or ritualized behavior that ignite communal passions. The Irish Congress of Trade Unions set up an antisectarian training unit called Counteract that does precisely this: creating a neutral environment and a mediated discussion forum to allow shop floor workers to discuss, for example, the wearing of football shirts, the acceptability or nonacceptability of particular lapel badges, or how the union should respond to external political developments. Resolutions of such discussions are sometimes framed around parity of esteem for both political cultures, but more often around agreements to remove all manifestations of communal allegiance. To provide another example, the Workers' Educational Association worked with the Transport Union to create a year-long educational program for bus drivers who were often exposed to violence on the streets and who wanted an opportunity to engage in a serious attempt to interrogate the policies of their own union. The program allowed them to meet with academics, politicians, and community leaders and through residentials and day schools to explore the roots of sectarian behavior and to develop appropriate responses.

Sectoral initiatives of this kind, like all developments within the workplace, operate within one bounded identity; the more difficult site is the local geographical community. While individuals move through as series of shifting and overlapping identities in their daily lives, it is their locality that provides the "anchor" identity. Where there are two communities with antithetical identities it might be expected that peace education would gravitate toward the coexistence end of the spectrum described by Maoz and that most work would be based on contact theory; in fact the movement is very much in the other direction toward more political work focused within single identity communities. Fitzduff (2002, p. 149) provides the rationale for working first within these monocultures:

> Communities, particularly those most ghettoized through history and locality frequently lack confidence, and can be too defensive and aggressive to engage in successful contact work. Single-identity work looks at ways to enable communities to look undefensively at the validity and worth of their own history and culture. It also includes work that enables groups to begin to identify issues on which they feel they can safely meet and co-operate with people from different communities.

This model is heavily indebted to ideas drawn from individual therapy, in particular the idea that dysfunctional relationships require self-analysis and personal growth before more healthy communications can develop. Transposed to community level, this suggests that the dialectical resolution of opposing identities can only be achieved when each community has developed such a confident sense of its identity that it can trade easily with others who, in a symmetrical way, have also learned to be more at ease with themselves. In practical terms, it has tended to mean Catholic nationalists immersing themselves more fully in courses in Irish language and culture, while Protestants unionists have veered off in a different direction to renew their sense of Britishness, with some inspired to construct a new Ulster-Scots language of their own. The danger of all this ethnic affirmation is that it acts to consolidate existing communalist identities, rather than to make them more fluid; and that rather than serving the creation of a shared identity, it acts to deepen the existing trenches. In the language of social capital theory, it deepens internal *bonding* (within communities already saturated with identity) without increasing *bridging* to other communities (Putnam, 2001).

No matter how a typology is constructed it has to leave a large space for the category of cultural diversity that has become the dominant discourse not only in Northern Ireland but in educational policy making across the globe. The promise everywhere is the same, and it is one that has particular appeal in situations of overt ethnic conflict. It is the promise that ethnic cultures can be drained of their toxicity and enjoyed as the enrichment of public space. The recent arrival in Northern Ireland of new immigrant groups from Nigeria, Portugal, and Eastern Europe adds some credibility to the idea of cultures slipping free from their political moorings, and of the hegemonic power of the two blocs being weakened by other ethnicities. Organizations such as Diversity 21 and the Community Relations Council organize festivals and fairs where Chinese dragons, Irish harpers, Orange drummers, and community circus performers come together in a cultural *mélange* that is intended to celebrate not difference—with its connotation of antagonism—but diversity. In Terry Eagleton's (2003) phrase, diversity becomes "post modernity's panacea for all ills." It is also an approach that lends itself easily to the practices of adult education where the traditional evening class or day school can be adapted to carry considerable freight through programs devoted to the study of other cultures and religions, and where study visits, readings, cookery sessions, and music performances can add a direct experiential element.

The openness of the multiculturalists approach is predicated upon a simple notion that all cultures are benign and equally deserving of respect. It is not a belief that withstands much scrutiny, and having been mauled for years by the right for "political correctness" it has now been subject to a much more searching critique by egalitarians like Brian Barry (2001) and Seyla Benhabib (2002). At the level of practice, however, the cultural diversity model continues to gain in strength, having been seized upon by a resurgent right both in America and in Britain as a justification for maintaining differentials in educational provision through the promotion of faith schools. What Levinson terms the "newly elevated demands of pluralism" (Levinson, 1999, p. 3) are deepening the divisions between communities, and this is as true of Northern Ireland as elsewhere. The Catholic Church, for example, no longer has to defend its separate schooling system in the old-fashioned language of "faith of our fathers"; it now speaks confidently of its separate schooling provision within the new cultural diversity paradigm that guides government policy.

### Evaluating Success and Offering Lessons for Elsewhere

As regards the lessons for other places, the necessary caution comes from Tolstoy's famous observation that, while all happy families are alike, unhappy families are all different in their unhappiness. And so it is with conflicted communities. Analyses of the Northern Ireland conflict have for many years been bedeviled by erroneous comparisons with other trouble spots like South Africa, the Middle East, Algeria, and the Balkans. It would be just as unhelpful to reverse this by insisting that the modalities of the peace process can be taken *mutatis mutandis* and as a template for elsewhere.

There are however reasonable questions to be asked: first, from those who invested political and financial capital in the process; and second, from those interested in conflict resolution and more particularly, given the context of this chapter, in peace education. To turn to the investors first: the biggest contribution to the peace process came from the European Union, which not only placed political faith in the process, but contributed two large-scale funding programs. The first of these, Peace 1, committed a total of 300 million euros to activities designed to build a sustainable peace. It ran through the period from 1996 to 2000 and was succeeded, like a movie sequel, by Peace 2, which in the tradition of movie sequels, had an even bigger budget: 500 million euros was committed in the period 2000 to 2006. Did the investment pay off? The official verdict from the EU is a positive one; in fact, so positive that additional monies have been forthcoming to allow conflict management skills developed in Northern Ireland to be brought to bear in other divided societies. At the macrolevel, it is easy to see why. However unsteady and zigzag its progress may have been, the peace process has taken Northern Ireland off the international casualty list and, by the standards of other peace processes (most notably the Middle East) it can be deemed to have been a success.

For those interested in peace education, the question has to be framed in a different way: how is the efficacy of different approaches to be judged? What makes for a successful peace education program? It is of course a notoriously difficult question, whatever the context. Out of approximately 300 items in a 1981–2000 literature survey on international peace education, only one-third (100) had elements of effective evaluation (Nevo and Brem, 2002). Of these, the majority (51) was deemed to be effective or partly effective. These verdicts themselves only serve to open up the larger question of what we understand by "effective." Here, a useful distinction is made by the American conflict resolution specialist Marc Howard Ross (2000) between the *internal* success of the project, and its impact on the *external* environment. Thus, a course designed to train 20 people in mediation techniques may be deemed successful if the organizers succeed in recruiting 20 participants who then go on to secure the relevant qualification; whether external conflicts are successfully mediated subsequent to the close of the program requires finer instruments of measurement. The bulk of evaluations that have taken place in Northern Ireland has concerned itself only with the internal measurements, though broader inferences have been drawn to connect these small drop-in-the bucket initiatives with larger movements in the peace process. The largest claims came in the wake of the Good Friday Agreement and the subsequent referenda held in the north and south to endorse the new peace accord. To some the political breakthrough was a triumph of people power or, in updated parlance, civil society, and those who had worked through the grassroots movements of community development and community education were the "unsung heroes" of the peace process (Cochrane, 2001). Other evaluations have been more cautious. In a mid-term evaluation by Cooper and Lybrand the problem is restated:

> It is apparent from our consultations with the responsible bodies, as well as the case studies undertaken, that there is no consensus on the precise means by which to ensure positive reconciliation aspects. A common theme in our workshops . . . is that there are considerable difficulties in defining peace and reconciliation impacts . . . Most groups surveyed (76%) claimed that their activities will have a direct impact on reconciliation. How this occurs is not something on which groups can be especially specific. (Coopers and Lybrand, 1998)

Some generalizations however must be risked. First, an obvious point: peace building is a long, slow process. That truism may win immediate accord, but its obviousness does not prevent it being ignored by the majority of evaluations conducted by funders. The problem of premature evaluation is inherent in all short-term funding arrangements; the really significant longer-term developments elude those whose attention is focused on short-term deliverables. A second point, related to this, also escapes evaluation processes based on a calculus of objectives, targets, and measurable outcomes, and it concerns the motivations of peace activists. In a series of eight interviews this author conducted with leading community relations workers in Northern Ireland, it was striking how little their efforts were guided by expectations of success. At the level of personal motivation. the key driver was a commitment to live out certain values which, in some cases but not all, was expressed as a form of witness to Christian belief. Even in its secular form this commitment was as a necessary part of the living out of a deeply held belief system. The question of whether the work achieved its goals was by no means thought to be unimportant, but more fundamental was a commitment to strive for goals even if they might never be realized. In some ultimate sense, peace processes everywhere rely upon this common core of idealism and the efforts of those who are prepared to push the rock up the hill time after time, knowing it will roll once again to the bottom.

Two other lessons suggest themselves. One is that whatever taxonomies are created of approaches to peace education the fine separation of different strands will not lead to the identification of the strongest: the strength comes from the braid. The weave in Northern Ireland has included both knowledge-based and affective learning strategies, education within single-identity communities and cross-community exchanges, strategies that focus on commonalities and strategies that dwell on differences. All of these are valid and all require cultivation. The success, or lack of success, will be determined crucially—and this final lesson is perhaps the most important of all—by the extent to which they help to build a common polity. This chapter has argued that multiculturalist strategies that aim for an innocent celebration of difference may actually deepen the fissures within a divided society, and that integration, not segregation, has to be the end goal. That goal cannot easily be reconciled with strategies that build upon difference. The support shown by the Blair government, for example to the development of faith schools (including those that teach creationism) appears particularly reckless in this regard. When disaffected Muslim youth rioted in the North of England in the 1990s, the official report chaired by Lord Ouseley on the rioting warned of the trouble ahead. His conclusion is worth quoting here, because it resonates not only with the Northern Ireland experience, but offers a warning to all societies that allow ethnic differences to eclipse shared civic identities:

> There are signs that communities are fragmenting along racial, cultural and faith lines. Segregation of schools is one indicator of this trend. Rather than seeing the emergence of a confident multicultural district, people's attitudes are hardening and difference is growing. (Ousley, 2001)

Events in Britain since that time have shown the prescience of the warning. It is not just in Britain however that communities are finding themselves drifting apart, and it is not just in Britain that ethnic and cultural ghettoes are under construction. Education, which ought to be offering a way out, is too often complicit with the divisions. Peace educators have not only to promote learning but, among the adult population, the unlearning of attitudes and prejudices. And we too, as educators have still much to learn, and have a pressing need to share our learning with others.

## References

Allport, G. (1954). *The nature of prejudice Reading*. Reading, MA: Addison-Wesley.
Area Development Management/Combat Poverty Agency (ADM/CPA) (2003). Building on peace: Supporting peace and reconciliation after 2006. Monaghan: ADM/CPA publications.
Bardon, J. (2001). *A history of Ulster*. Belfast: Blackstaff Press.
Barry, B. (2001). *Culture and equality*. Cambridge: Polity.
Benhabib, S. (2002). *The claims of culture*. Woodstock: Princeton University Press.
Bush, K.D. and Saltarelli, D. (2000). *The two faces of education in ethnic conflict: Towards a peacebuilding education for children*. Florence: UNICEF, Innocenti Research Centre.
Cochrane, F. (2001). Unsung heroes: The role of peace and reconciliation organizations in the Northern Ireland conflict in McGarry, J. (Ed.), *Northern Ireland and the divided world*. Oxford: Oxford University Press.
Cochrane, F. and Dunn, S. (2002). *People power: The role of the voluntary and community sector in the Northern Ireland conflict*. Cork: Cork University Press.
Connolly, P. (2003). The development of young children's ethnic identities: Implications for early years practice in Vincent, C. (Ed.), *Social justice education and identity*. London: RoutledgeFalmer.
Coopers and Lybrand. (1998). *Mid-term evaluation report of the peace and reconciliation programme*. Belfast: Coopers and Lybrand.
Department of Education for Northern Ireland (1998). *Lifelong learning—A new learning culture for all*. Belfast: Her Majesty's Stationery Office.
Department for Employment and Learning (2004). *Further education means business*. Belfast: Her Majesty's Stationery Office.
Eagleton, T. (2003). The roots of terror, Guardian, September 6, 2003, p. 21.
Equality Commission (2001). Monitoring report 11: A profile of the Northern Ireland workforce, 2000. Belfast.
Fitzduff, M. (2002). *Beyond violence: Conflict resolution process in Northern Ireland*. Tokyo: United Nations University Press.
Foster, R.F. (1990). *Modern Ireland 1600–1972*. London: Allen Lane.
Gray, J. (1995). *Enlightenment's wake*. London: Routledge.
Gutman, A. (1995). Civic education and social diversity. *Ethics*, 105 (3): 557–579.
Hargie, O. and Dickson, D. (2003). *Researching the troubles: Social science perspectives on the Northern Ireland conflict*. Edinburgh: Mainstream Publishing.
Jennings, B. (1979). *Knowledge is power: A short history of the WEA*. Newland Papers, Hull: University of Hull.
Kember, D. Kwan K-P., and Ledesma, J. (2001). Conceptions of good teaching and how they influence the way adult learners and school-leavers are taught. *International Journal of Lifelong Education*, 20 (5): 393–404.
Kymlicka, W. (1995). *Multicultural citizenship*. Oxford: Clarendon Press.
Levinson, M. (1999). *The demands of liberal education*. Oxford: Oxford University Press.
Longley, E. (1993). In Pollak, A. (Ed.), *A citizen's inquiry*. Dublin: Lilliput Press.
Macintyre, A. (1985). After virtue London: Duckworth.
Maoz, I. (2002). Conceptual mapping and evaluation of peace education. In Salomon G. and Nevo, B. (Eds.), *Peace education*. London: Lawrence Erblaum Publishers.
Nevo, B. and Brem, I. (2002). Peace education programs and the evaluation of their effectiveness. In Salomon, G. and Nevo, B. (Eds.), *Peace education* (pp. 271–283). London: Lawrence Erblaum Publishers.
Northern Ireland Council for Integrated Education (undated). *Integrating through understanding Belfast*. NICIE Resource Pack (n. p.).
Ousley, H. (2001). *Community pride not prejudice: Making diversity work in Bradford*. Bradford City Council.

Pettigrew, T. (1998). Intergroup contact theory. *Annual Review of Psychology*, 49: 65–85.
Pollak, A. (Ed)., (1993). *A citizen's inquiry: The Opsahl report on Northern Ireland.* Dublin: Lilliput Press.
Putnam, R. (2001). *Bowling Alone: The collapse and revival of American community.* New York: Simon and Schuster.
Ross, M.H. (2000). Creating the conditions for peace-making. *Ethnic and Racial Studies*, 23 (6): 1002–1034.
Ryan, S. (1995). Transforming violent intercommunal conflict as quoted in Cochrane and Dunn, *People power*, p. 78.
Shirlow, P., Murtagh, B. et al. (2001). *Spaces of fear: Measuring and modelling chill factors in Belfast.* Coleraine: University of Ulster Research Report.
Smith, A. and Vaux, T. (2003) *Education, Conflict and International Development.* London: Department for International Development.
Tomasevski, K. (2003). *A report on the mission to Northern Ireland November 24 to December 1, 2002 by the UN special rapporteur on the right to education.* Geneva: United Nations.
Trew, K. (1976). Catholic-Protestant contact in Northern Ireland. In Hewston, M. and Brown, R. (Eds.), *Contact and conflict in intergroup encounters.* Oxford: Butterworth-Heinemann.
Vincent, J. (1993). Ethnicity and the state in Northern Ireland. In Toland, J.D. (Ed.), *Ethnicity and the state.* London: Transaction.

# Section V
# Teacher Education

# Chapter Fifteen

# The Reconstruction of the Teacher's Psyche in Rwanda: The Theory and Practice of Peace Education at Kigali Institute of Education

*George K. Njoroge*

### Introduction

Located in central Africa, Rwanda measures 26,338 square kilometers and has a population of 8 million people. The country has three major population groups: Tutsi, Hutu, and Twa. Rwanda came into the limelight negatively during the 1994 genocide when 1 million people mainly Tutsi and moderate Hutu people were killed in just 100 days. The genocide caused a multiplicity of problems. Among the strategies used to reconstruct the country was the use of education to develop decimated human resources. While human resources function in many fields, quality teacher education emerges as an imperative preventative measure against future genocide.

The reconstruction of the teacher's psyche in Rwanda can only be understood within a context of discrimination and violence that has characterized Rwandan society since colonialism. Ill-prepared and indoctrinated teachers cannot be positive change agents. A country that desires to live beyond tragedy requires that teachers be knowledgeable and equipped with the skills necessary to respond to challenges that may have been the cause of violence and dehumanization. The attempt at reconstruction of the teacher-learners' psyche at KIE (Kigali Institute of Education) is based on the fact that teachers are among those accused of committing genocide in 1994 in Rwanda. However, they are today expected to be positive change agents. They are expected to promote among Rwandans solidarity, respect, and care for others alongside other, fundamental values.

### Construction of a Dehumanized Self through Education in Rwanda

Access to secondary school in colonial Rwanda was highly selective and subject to systematic anti-Hutu discrimination. The Catholic Church and the colonial administration favored the Tutsi. Like elsewhere in Africa, colonial and missionary education in Rwanda sought to reinforce the divide and rule philosophy. Thus, in schools the differences between Hutu and Tutsi pupils were emphasized. Pupils were categorized in and out of school (Rutayisire et al., 2004). This process of differentiation was later perfected in postcolonial education by means of ethnically defined identification files for the pupils, biased examinations, violent forms of punishments, biased history and civics curriculum and a quota system that defined access to education on the basis of ethnic, regional, and gender criteria.

During the first postcolonial government/republic under the leadership of Gregorie Kayibanda, the importance of schools in propagating hate ideology emerged through the establishment of a movement at the secondary school and university level aimed at reducing or eliminating Tutsi presence from schools and from the private sector. The Tutsi were reproached for previously monopolizing these two areas where indeed they had exerted considerable influence mainly because of earlier discrimination that had favored the Tutsi. Violence was often used to flush out Tutsi teachers and students from school through "Public Salvation Committees" (Gasana et al., 1999). Tutsi were socially constructed as public enemies and the public was essentially Hutu.

In 1973, the second republic, formed after the overthrow of Gregorie Kayibanda by Juvenal Habyarimana, was characterized by discrimination. Habyarimana implemented a policy of regional and ethnic equilibrium where Hutu who came from the south, Gregorie Kayibanda's region, as well as Tutsi were discriminated against. Habyarimana's policy was a process of sectarian exclusion (Gasana et al., 1999).

Hatred between Hutu and Tutsi was built over a long period through teachers' classroom practices. Ideology that glorified irreconcilable differences portrayed all Tutsi as perpetrators of historical injustice toward the Hutu and history taught as a subject in school passed on this message:

> The relentless propaganda, underpinned by an educational system that divided the Rwandese people and which reinforced many dangerous historical myths and stereotypes, created a . . . consciousness in the Hutu community that the extremists could play upon to encourage people to kill their neighbors and colleagues. (African Rights, 1995)

The role of education in fomenting hatred and conflict and the actualization of this ideology in practical terms can be attested by the educator's participation in genocide. It is noted that

> [m]ore than any profession, academics, teachers, school inspectors and directors of schools, including primary schools, participated actively in genocide. Throughout Rwanda, they helped organize the killing squads and took a lead role in the hunt for victims and in carrying out the massacres. In some cases, teachers murdered children who attended their own schools or even learned in their own classes. Many academics and teachers are responsible for the death of their colleagues. (African Rights, 1995)

What the above shows is that education can be used as the foundation for doom or for hope depending on the values upon which such education is founded and the objectives it strives to achieve. Teachers can mold or destroy their learners. The education given before 1994 helped construct a totally dehumanized being bereft of any feeling and able to kill without any mercy, conscience, or moral guilt.

The school in postgenocide Rwanda has been defined anew as a tool for positive social transformation. This role must be understood in light of the country's new philosophy that perceives all Rwandans as citizens. This transformation requires that all Rwandan's view of the other be reconstructed accordingly. The need for this is based on one of the consequences of genocide where the Tutsi are no longer seen as refugees, but as victims and the Hutu seen as genociders. Continued perpetuation of these identifications negates the desire and endeavor for unity and reconciliation as well as peace. Moreover, it dichotomizes the citizens into polar opposites. Constructed as victims, the Tutsi are dehumanized and denied their ability to transcend the negative experiences they have undergone. As victims their psyche is captured by those who dehumanized them. It becomes difficult for them to reconstruct their lives and to affirm their dignity. On the other side of the divide, constructed as genociders, the Hutu in toto are dehumanized without blame being apportioned to those particular individuals who committed the genocide. The guilt of those who committed genocide enslaves them and denies them their ability to affirm their personhood and continue with their lives. Gasana et al. (1999) note:

> Since the genocide, there exists a deep ethnic polarization within Rwandan society; this is not generally manifested in open tension between the various group members, but rather by a feeling of mistrust. Relationships are superficial and remain essentially utilitarian. In the perception of the individual there is no basis for trust outside one's ethnic group.

The search for peace entails transcendence of the perception of the other as the enemy. The relationship between the two groups has always been characterized by schisms. Thus, it requires courage to extend one's hand to the other—groups who had previously been perceived as the negation of oneself. If the Rwandan society wishes to kindle hope for future generations of Rwandans that they can live together, they must accept the negative history of genocide. In the words of Linden:

> "Post-conflict" peace building, . . . requires giving attention to the burden of history and the construction of socially unifying stories about the past, large enough to encompass the pain of conflict and killing, capable of gaining the assent of contending parties, yet not new ideologies of power in themselves. (Linden, 1999)

The call to schools to help in the reconstruction of the psyche of Rwandans poses a great challenge. Schools themselves as well as all those involved in the school must be transformed. Throughout this transformation, one of the most important players is the teacher. He or she must be the first to be transformed. Good programs fail if teachers cannot translate the programs into practice as required or if they act as saboteurs because of their own prejudices.

## Reconstruction of the Psyche

Hinchcliff (1996) argues that in order to know where we want to go, we need to recognize which myths prevail in our society and which alternative ideals and values are needed in order to achieve a metanoia, a new mind-set. This is an expression of the reconstruction of the psyche because it entails the reorientation of a person's way of viewing and interrogating reality. It does not just happen and must be founded on certain fundamental human values. One of the roles of education is to ingrain such values. Toh Swee Hin (1997) supports this position emphasizing the critical importance of the reconstruction of the mind for peace education. He notes that if education fails to move minds as well as hearts and spirits for personal and social action in peace building, it ends up being emasculated and co-opted by forces interested in preserving the status quo.

Reconstruction of the psyche is multifaceted. It involves the reorientation of a person's perception of relationships with others and the roles that have hitherto been assigned to him/her on the basis of dehumanizing criteria. It strives to capture and ingrain an attitude and practice of recognition for others' humanity and right to search for dignity and personhood. It aims to commend differences because, after all, human beings are human because they are different and the world would be boring and unproductive if humans were all the same. Fundamental human differences can be used in the construction of unifying memories that act as a key reminder of what the inability to appreciate the unifying potential of human diversity can mean to a country.

Reconstructionism in Rwanda endeavors to create a present and future free from the evils of discrimination and dehumanization. However, this recreation does not pretend to ignore the past. The past is enshrined in order to guide and remind people of the tragedy that can befall human beings if they forget the past. Reconstruction is preferred because it reorients rather than erases the psyche so that the past can be used as a reference bank should the person veer from that which is expected of him/her.

The main premise on which reconstruction is based is that people can change. As such, we have to see the potential in what may be assumed to be a hopeless situation. The teacher, as the implementer of reorienting education, works from this hope. Indeed, experience has shown that

> Humanity promotes violence but is also capable of promoting peace. However, to help society fully comprehend this nature of humanity and work towards restoring the hope that it is possible to exploit the good side of this humanity, the challenge is thrown to the teacher. The teacher is therefore expected to scale the ladder with his/her charges in the search for a peaceful Rwandan society. The youth are expected to transcend the pre-judgmental cave of superior or human 'US' and inferior or inhuman 'THEM' familiar certainties that have beleaguered the past history of Rwanda. (Njoroge, 2002)

Teachers, like all people, cannot be transformative if they are not liberated from the prejudgmental cave of hate, discrimination, and ordering of other persons as superior or inferior. Liberation entails dealing with the psyche, or what John Shotter (1997) calls an entity with constantly contested and shifting boundaries; something we can recollect in one way one day, and in another the next.

Today education emphasizes the re-orientation of its recipients to be rational and creative interrogators of their reality. It is a concerted endeavor to enable their psyche to resist anaesthetization and dehumanization that has denied them their humanity in the past. Indeed, postgenocide education policy makers believe that the essence of education is to recreate in young people the values that have been eroded over the course of the country's ... history (Republic of Rwanda, Study of the Education Sector in Rwanda, 1998, p. 23).

The need for life skills as a new addition to the curriculum aims to help learners to become self-directed rather than other-directed. According to R. Murenzi, the curriculum failed the nation in 1994. He asserts that

> The curriculum was both silent in areas where it should have been eloquent and eloquent where it should have been silent. For instance, there was too much about human differences and too little about human similarities . . . too much about theory and too little about practice. (Murenzi, 2002)

This is a clear statement of warning that postgenocide education must focus on similarities and potentials that are fruitful to the individual and the country.

Education needs to be a manifestation of promulgated policies and objectives that enhance the development of individuals in a full sense. The most difficult task is putting into practice the promulgated polices. Their implementation and practice involves individuals who have been members of a dichotomized society, used to mistrust and hate. Transforming individuals to fully embrace a new paradigm shift is not an easy task. And so every institution gets into a kind of quagmire even as it goes about implementing educational policy.

Educating teachers to become re-creators of a new Rwanda requires putting into context their social milieu. These teachers have long been part of a society characterized by schisms. Today they are expected to steer change in their classrooms and schools. They can only do this if they understand clearly the context in which they operate. It is for this reason that (Woods, 2002) argues that reform in the education of teachers is of paramount importance in Rwanda today. He suggests that Rwandan teachers be trained, educated, and supported to adopt a teaching style that differs from the one that they themselves experienced as students.

The call by Woods is supported by Byuma, former Rector of KIE, who asserts that educators should open teachers' minds and give them a sense of purpose and direction. This process will help teachers to teach and interpret the curriculum so as to ensure the production of healthy, conscious citizens who understand and value peace, democracy, and unity (Byuma, 2002). Education of teachers therefore entails helping them to examine themselves as individuals in order to recognize and address their own inadequacies. In particular, this entails teaching teachers to be saboteurs of what has always been taken to be the norm in Rwanda—that is the existence of an unconscious denigrated self who is superior to the other and a conscious inferior self depending on one who is in power.

This fundamentally means that teachers must be educated in such a way that they transfer their humanity to their classrooms. Rutayisire (2004) opines that this calls for the touching of the heart, mind and soul of the teacher in order to create a new generation of young people who hopefully will be the future of Rwanda. As such, the classroom must be recreated into a site for growth and affirmation of life. Moreover, relations and knowledge constructed must not be for the anaesthetization of any participant. A galaxy of diverse cultures and interpretation of reality should be the bedrock for inventiveness and creativity. All this requires a new orientation for the teachers, in other words a focus on similarities and the potentials of cointerrogation emphasizing what it means to be Rwandan rather than a Tutsi, Hutu, or Twa.

## Teacher Development at Kigali Institute of Education (KIE)

The government established KIE in 1999 with the mission of making it a center for excellence in teacher education. The vision behind KIE was the nurturing of quality in teacher development for all levels of the education system (KIE, 2000). There is no gainsaying the fact that the long-term risk of a lack of qualified teachers is the deterioration of academic standards and the country's development both in terms of physical dimensions and in human dimensions (Maritim and Katia, 1998). For too long Rwandan teachers, like everyone else, were steeped in ignorance as observed earlier. They were robotic propagators of discrimination, stereotypes and prejudice as well as hate. KIE is expected to help change this by being one of the institutions to adopt a new thinking and philosophy called social reconstructionism.

The KIE teacher-learner takes two teaching subjects alongside education. For example, students may take mathematics, physics, and education. During the beginning of the third, fourth and final years of study, they are also expected to teach. The idea of the teacher-learner taking two teaching subjects as well as the school practice requirement for recognition as a professionally qualified teacher is an innovation in teacher education in Rwanda. This is meant to ensure availability of teachers in all areas of the curriculum as well as teachers who are aware of practical and existential dynamics at play in the Rwandan classrooms. The faculty of education empowers teacher-learners with methods for dealing with multifarious classroom and school challenges.

The innovations were meant to address inadequacies present in previous teacher education programs. In the pregenocide programs, a teacher could be anybody who had a diploma or degree in either science or art subjects. They lacked basic, professional and other fundamental skills required for being a teacher. This may be one of the reasons "teachers" easily fell to hate ideology and propagated it with abandon and enthusiasm. They had no conception of what it means to be a teacher who is a critical and creative leader. A professional teacher is ideally a critical, creative, and transformative agent of hope and life.

Developing highly qualified teachers is an onerous task and enormous challenge. KIE aims to educate qualified teachers not just by awarding them certificates, but also by monitoring the quality of teacher output in schools and in the community at large. What teachers do for the self and for Rwandan society—in line with the national philosophy—is of fundamental importance. Teachers are expected to be agents of the reconstruction of the self and the Rwandan psyche in general.

KIE-educated teachers are envisaged as role models. They shake the society from ambivalent assumptions about the irrelevance of events, yet they have an impact on people's everyday reality. The teacher is expected to deconstruct the vicious circle of ignorance and violence and liberate the mind into knowledge, skills and change in attitudes whose power will shape the nature of reality in interpersonal and ecological relations that will be amenable to people living peacefully and fruitfully in Rwanda (Njoroge, 2002).

In Rwanda today, those involved in education have a chance to revisit their orientations in a manner that can result in a normative or ethical being. In this context, education should promote the capacity for connectedness and solidarity that underlies ethical desires in Rwandan society. It is an engagement with personal social responsibility. It is not just a question of changing mind-sets but rather a living process that responds to everyday challenges. It is an interplay between values, knowledge, attitude change, as well as skills and their practice.

## Philosophical and Pedagogical Underpinnings of KIE's Life Skills Program

The spirit behind KIE's life skills program contends that human challenges can be fruitfully addressed only through a dialogical engagement between different players. This engagement acknowledges the humanity of all those involved despite an ever-challenging terrain of life. To some extent the philosophy of Paulo Freire, a Brazilian philosopher, informs the stance taken by KIE. Students—envisioned as transformative agents in a new Rwanda—are expected to transcend the boundedness of their past, anaesthetizing reality as well as the past of others, with whom they will teach and learn (Freire, 1998, pp. 100–101). KIE's life skills program attempts to actualize the notion that teachers can be agents for hope and peace in Rwanda.

The link between life skills and holistic learning and living is well stated in the KIE life skills development report (KIE, 2003). Indeed, holistic learning and living are about developing knowledge of and approaches toward the integration of life-enhancing attitudes, universal values such as peace, and creative holistic techniques. In giving the teacher-learner the skills, it is hoped that they will practice them beyond their life at KIE. Teachers are important change agents but can only give to their students the knowledge and skills that they themselves know about and practice. The life skills program is an attempt to link education with real life challenges in Rwandan schools (Njoroge, 2004).

Life skills include personal and social skills that enable learners to function with dignity, to respect others as equals and to honor the community upon which the learners' existence is incumbent (KIE, 2003). Life skills are abilities which help individuals to be successful in living a productive and satisfying life. The goal of teaching life skills in peace education is to provide appropriate opportunities for students to experience life fully, while conscious that the "I" that defines the self cannot be realized without the "WE" that is the community of which the "I" is a part.

In view of this, peace education can be defined as a holistic process that gives people the skills and knowledge necessary to develop critical thinking and participatory learning around a wide range of issues pertaining to the lived reality in Rwanda (KIE, 2003). Peace education strives to promote tolerance, national unity, and knowledge of the self while guarding against genocide ideologies, prejudice, and stereotypes along with other negative values that have militated against coexistence.

## Challenges to Peace Education at KIE

Various challenges that militate against peace education can be identified at Kigali Institute of Education and generalized to Rwandan society as a whole. Among these challenges are stereotypes and prejudices that are viewed as nugatory to peace. These stereotypes and prejudices are related to:

- Francophone-Anglophone dichotomy
- Flight from reality
- Question of origin.

## Francophone-Anglophone Dichotomy

Before 1994, the languages in use in Rwanda were Kinyarwanda as a national language, Swahili as a regional lingua franca and French as the only official language and the medium of instruction in schools. Today, Rwanda has a trilingual policy that emphasizes Kinyarwanda, French, and English. The adoption of this policy stems from the fact that, after the genocide, Rwandans returned to Rwanda from all parts of the region and from Europe. This necessitated a language policy that would address the diversity of cultures and identities Rwandans had constructed in the countries to which they had fled, as refugees since the pogrom of the Tutsi in 1959 and in postindependent Rwanda. The English and French languages become new labeling tools among Rwandans perpetuating the "US" versus "THEM" dichotomy that challenges the teaching and practice of peace education skills.

In most cases, when teacher-learners are asked about their identity, they define themselves as either Anglophones or Francophones. This is a manifestation of the negating ideology that has been entrenched into their psyches over the course of time. They rarely define themselves as Rwandans. They use languages to define their nationality and by defining themselves as either Anglophone or Francophone they manifest what Paulo Freire says is an absence of citizenship (Freire, 1998).

In part, language determines behavior and consequently identity. For example, lack of time consciousness and absenteeism are identifiable marks of the school's English-speaking staff work ethic. In other words, French-speaking staff is thought to have a superior work ethic. In addition, the English-speaking pupils are said to be low achievers when compared with their French counterparts (African Rights, 2001). This schism between groups in schools is expressed in stereotypes that dichotomize Rwandans, not in previously used terms such as Hutu or Tutsi, but as Anglophones or Francophones. The use of these labels poses a challenge that the peace education life skills program cannot ignore.

## Flight from Reality

Today, people in Rwanda feel uncomfortable using the terms Tutsi, Hutu, and Twa in their daily conversations. However, the theory and practice of peace education skills cannot ignore the subject of ethnicity. In the classrooms, all ethnic groups are represented. Therefore, problems arising around ethnicity cannot just be wished away with the hope that they will disappear. Accordingly, teachers are challenged to teach and conduct themselves in a manner that does not favor or stereotype one ethnic group or the other. Moreover, they are encouraged to address this taboo area as a matter of urgency. This is important if Rwandans want to find solutions to social divisions that require an acknowledgment of ethnic categorizations, ethnicity, and ethnic groups. Once such categorizations are acknowledged, it is easy to express a consensus as to the truth about them as they are experienced and defined (Rutayisire et al., 2004).

This is all the more important if one considers that teachers have to participate in transforming the society yet they also fall into one of these ethnic groups. How do they perceive their colleagues and the students they teach? Are they immune to the schisms that have previously affected Rwandan society? Rwanda cannot afford to ignore the challenge of ethnicity and its effect on social relations.

## The Question of Origin

According to Gyezaho (2004), Rwandans classify each other as *Abasopecya, Abasaja, Abajepe, and Abadubayi* among other stereotypes. These stereotypical labels can be explained as follows:

- Abasopecya—this is a Rwandan who is presumed to have been born in Rwanda and lacks exposure to the outside world. Sopecya comes from a petrol station in Kigali the capital city of Rwanda that never closed its services during the 1994 war and genocide. Thus, it is opined that Abasopecya are those Rwandans who survived war and genocide. They are said to be dangerous and hardhearted and ready to accept whatever comes their way.
- Abasaja—this is a Rwandan who was in exile in Uganda. The word comes from the Ugandan term msaja, which means man. Those from Uganda are said to be real men. They are considered to be proud of their background and are said to be "hardworking."
- Abajepe—this is a Rwandan who was exiled in Burundi. The word comes from corrupted (Guarde Presidentiale). Because of proximity to power, Abajepe sample the good life and other amenities that go with power. They are therefore said to believe in leading a soft life that comes not because of their efforts but because of opportunities. They are said to be materialists and opportunists.
- Abadubayi—this is a Rwandan exiled to the Democratic Republic of Congo. Dubayi comes from the word Dubai, the duty free capital in the Middle East known for its second-hand car exports to Africa. By appearance, the exports glitter. In actuality, they are rotten inside, a manifestation of greed of those who export and those who import. Thus, Abadubayi are said to be greedy despite their outer demeanor of kindness and hospitality. They are symbols of what is wrong with the Democratic Republic of Rwanda (DRC).

Though it may seem that there are many categories in the stereotypes presented, the groups are basically two: refugees and those who remained in Rwanda. This dichotomy has social and political ramifications that Rwanda today is striving hard to deal with, through initiatives that incorporate the inculcation of positive human values. The existence of these stereotypes is a manifestation of a clash of worldviews, interpretation of lived reality and how one constructs personal identity vis-à-vis the other.

### An Attempt to Deal with Stereotypes at KIE: The Case of a Philosophy of Education Course Assignment

In 2004, the above-mentioned stereotypes served as a reference point for a group assignment in a philosophy of education course. The objective was to test how teacher–learners could apply their knowledge of ethics of care as promulgated by Noddings and Diller (Noddings, 1984, 1992; Diller, 1996) to their contextual reality at KIE. The application was guided by one of the objectives of education, which states, in part, that the purpose of education in Rwanda is *to educate a free citizen who is liberated from all kinds of discrimination, exclusion and favoritism.* In light of this stated philosophy, and with reference to the principles of ethics of care, teacher-learners were required—among other things—to

1. explain the ontologies manifested by the existence of stereotypes as well as the relational ideals that were violated and
2. describe the changes that they had undergone as a group since they started the assignment as persons who had not shared an assignment before or had different views of each other.

Normally, when students at KIE are asked to form discussion or other assignment groups, they constitute them according to subject combinations or according to their knowledge of each other. Given the history of Rwanda, there is a tendency to work with those with the same origin, language, or interpretation of reality. In this course however, 26 groups each composed of 14 teacher-learners were constituted. Teacher-learners from the faculties of science, arts, and social sciences were mixed together. This was meant to reorient the teacher-learners away from their usual practice.

The challenge to the teacher-learners to work in newly formed groups was initially met with great opposition. Many were afraid because they did not know the other teacher-learners. Moreover, they had stereotypical ideas and prejudices against the others.

In response to their uneasiness about exploring new terrain, teacher-learners were reminded that they had just learned about ethics of care and that they were to apply the principles that included encounter, coexploration, cooperation, and dialogue and so on. Thus, part of the assignment was to discover how to find the distant other, the person they did not know but who was supposed to be a member of their group. While this new reality was frightening, it was meant to put learned theory into practice.

## Constructed Transformations after the Course Assignment

When teacher-learners were asked to describe some of the changes they had undergone as a group since embarking on the assignment, they indicated that they had focused on previously held stereotypes and prejudices as well as the problems that they faced in accomplishing the task at hand. The comments shown here illustrate how one can draw fundamental lessons for dealing with stereotypes and prejudices through peace education in postconflict societies.

The comments given by the teacher-learners indicate that the assignment, in some way, helped bridge the stereotypes between arts and science students. Those who thought negatively about the other group discovered the truth about them. According to some teacher-learners:

> The assignment helped group members to eliminate some prejudices and stereotypes that were among KIE students. The students had misjudged and made biased statements against their fellow students. Arts and social science students were thought to be lazy and thus studied easy subject combinations compared to science students. The arts students on the other hand thought that science students could not be able to argue out

philosophical questions but this was proved wrong because their presence and participation in the group assignment was acknowledged and praised by all. Meeting and working together in these new groups made us realize the potential that each one of us have if given the chance. (KIE-Group 19 and 24, 2004)

With reference to ontological relationships regarding stereotypes emanating from places of origin, some teachers opined that the groupings were nonessential and/or not helpful. After the group work, students felt that they had come to know one another and realized that they were all KIE students and, what is more, Rwandans for that matter. Their collaboration was bound to grow. (KIE-Group 15, 2004). The argument by this group is simplistic because it wishes away factors that led to the stereotypes in the first place. However, the recognition that all are Rwandans is important and represents a first step in a creative and critical endeavor to address some of the challenges that may be taken for granted.

Friendship is important and it enhances the capacity for human beings to relate. It cannot be established if there is no close encounter between persons. It is through this encounter that persons are able to seize, interrogate each other and then create a bond that seals friendship. It must be based on the understanding that all human beings are capable of communicating with each other. Within this context, we can understand the teacher-learners who asserted that through the assignment they found new friends. They had not even considered them before but the encounter helped them realize that the "others" were good people (KIE-Group 14, 2004).

With regard to communication, at first one group thought that the Anglophone-Francophone dichotomy would be a hindrance. However, it melted away when they started the assignment. They found they could communicate. This was possible because of patience and tolerance (KIE-Group 10, 2004). This observation shows the entrenchment of the Anglophone-Francophone schism. Though Kinyarwanda is the lingua franca, without apology students perceive themselves as either Francophone or Anglophone.

The task given to the students was not an easy one and was intentionally designed to give them a chance to experience the difficulties that caring for the other entails. This is even more difficult when one engages in a task which, without the other, cannot be accomplished successfully. According to some of the students, the greatest difficulty was to find each other. Patience, tolerance, and sacrifice were mentioned as virtues that helped them to accomplish the task successfully and on time (KIE-Group 22, 2004).

The group assignment helped to awaken in the participants personal responsibility in what they can do without dependency on others as well as collectiveness in working with others. When students are given group assignments, a common phenomenon is that the tasks end up being done by those identified as brighter than their colleagues. The students had this to say about their experiences in the new arrangement in comparison with the previous method of group formation:

> Because of familiarity, we had been used to few students in the previous groups doing the assignment and at the end of the day, the others signing that they had participated in the assignments. The new arrangement could not allow this and everybody had to participate because it was a new encounter. It was a case of encountering others at close proximity and sharing with them. (KIE-Group 6, 2004)

We may observe that in the new groups, the teacher-learners were not familiar with each other. This may have created fear that nonparticipation could have its attendant problems. However, it could also be taken to be an effort at breaking long-held traditions.

## Lessons from the KIE Experience

We cannot rule out the fact that teacher-learners may have cooperated during the assignment for utilitarian purposes (i.e., they were being awarded marks). However, the experience itself, of working and encountering others, is a first step toward the practice of skills in peace education. Peace is not achieved overnight. It is a long and tedious process. The fact that the teacher-learners made positive references to their changed perception of the others gives credence to the belief that human beings can change though it takes greater effort than one course assignment encounter.

In peace education, pedagogy matters. It is not just a delivery of skills but a living process that requires all those involved to make everyday encounters with the other. This goes beyond the mere acknowledgment that the other exists. It requires a deep search for the humanity of those involved through interrogative encounters. This cannot happen until one person has communicated with the other enabling one to begin to appreciate that indeed the other is as human as oneself. Prejudices and stereotypes that may cloud one's thinking need to be cleared before making the move to relate to the other. Indeed, it is a question of working toward the promotion of relational encounters that each one is able to contribute. This is emphasized by Gergen (1995) when he asserts:

> Societal transformation is not a matter of changing minds and hearts, political values or the sense of the good. Rather, transformation will require unleashing the positive potential inherent in relational processes. In effect, we must locate a range of relational forms that enable collective transformation as opposed to alienated dissociation. (Gergen, 1995)

Any endeavor to transform a people needs to take into consideration the relational dynamics that steer their lives. For peace to exist in the schools, for example, the way the teacher relates to the students and the way students relate to themselves is a matter of utmost importance. However, this cannot be achieved if students are not helped to acquire a self-focused critical posture and are taught not to focus on the other who is thought to be different in a negative sense. In focusing on the other to whom one has already allotted stereotypical and prejudicial values, one becomes a slave and lessens his/her own capacity to fruitfully affirm the self.

Peace education needs respect for the other and cannot thrive where there is no compassion. But, compassion itself must be built through continuous effort to promote the other as human. This can only be done where encounters are based on sharing and solidarity that starts from the precept that the other is as human as oneself. Indeed, any pedagogy concerned with peace has to address the issue of not how people think and cognitively know the world, but also the extent to which they develop their capacity for feeling, empathy, and emotional connection to themselves and to others (Shapiro, 2001).

There is a need to engage in pedagogy that bridges the gap between thinking and lived reality. Students need to value solidarity with the other as they try to interrogate the world. There is no gainsaying that this calls for construction of identity that is in constant conversation with the world in which the students live. Ignoring the dynamics of this reality will not help foster peace because the student will always be in competition with others. Shapiro decries the entrenchment of pedagogy of dispossession in schools in modern times when she observes that in schools, students learn to value competition over community, self over other and disconnected information over connected knowing (Shapiro, 2001).

In a post conflict country, the curriculum for peace education must be promulgated with the necessary change in pedagogy. Introducing a curriculum without the teachers undergoing a change in their total psyche with regard to teaching-learning and the consequent practice of skills learned is a sure recipe for failure. Peace education requires that teachers be creative and practical at the same time. They need to engage in pedagogy that affirms the liberation not only of the students but also of the self. One cannot help reorient the student's thinking or psyche without practicing what one preaches.

In order to deal with stereotypes and prejudices that may inform the teacher practices in the classroom, those concerned must be open to what happens in the school and society in general. The teacher-learners, for example, need to know about stereotypes and prejudices and endeavor to change them practically through a caring pedagogy.

An education that is for the affirmation of the rights of all requires transcending the inclination of both potential citizens and educators to define themselves primarily in terms of group loyalties and identities (Nussabaum, 1997). This holds true for teacher-learners and their lecturers. The teacher-learners as potential peace builders and their lecturers need to transcend the history of discrimination and stereotypes that have typified their relations. They must reproach themselves and transcend the constructed identities that have tended to glorify differences and ethnicity resulting in a constant feeling of insecurity and fear of the other.

A thorough and clear articulation of issues that threaten peace in society while giving students a chance to air their views is a fundamental strategy in a caring pedagogy. The reorientation of the psyche is a process of empowering teacher-learners to transcend their prejudices and ethnocentric views on what counts to be a person in society. It is a move to change oneself into a liberated and emancipated being who perceives the self as a person who has equal rights as the other in the search for dignity, personhood, and identity.

## Conclusion

Classes or lecture rooms should be places for the celebration of differences, places where diversity and its potential for peace, realization of dignity and personhood for all the people is recognized, nurtured and lived in full. They should not be places for division and dehumanization but places for reaching out to the human other.

Teacher-learners who, after graduation, are meant to reconstruct the psyche of Rwandan society must emerge from their education as saboteurs of ethnicity, hate, discrimination, stereotypes and prejudices—the hegemonic plagues that have characterized Rwandan society for years.

The aim of educating teachers in Rwanda is to make them effective transformative agents. However, they cannot be effective without the necessary skills that are contextual to the promulgated philosophy. Teachers in Rwanda have to be guided in all that they do by the philosophy of social reconstruction. In order for this philosophy to be actualized, institutions must creatively design methods and means of addressing various challenges encountered. While all aspects of the reconstruction of the psyche are not covered in this chapter, life skills in peace education has been taken as a case study in an effort to address the challenge of stereotypes and prejudice while basing it on pedagogy of care in the training of teacher-learners at KIE.

## References

African Rights (1995). *Rwanda not so innocent: When women become killers.* London: African Rights.
———. (2001). *The heart of education: Assessing human rights in Rwanda's secondary schools.* Kigali: African Rights.
Byuma, I. (2002). *Curriculum and teacher training Republic of Rwanda, Report of the national curriculum conference—Curriculum in the service of national development: What skills do our children need?* Kigali: National Curriculum Development Centre.
Diller, A. (1996). Pluralisms for education: An ethics of care. Retrieved from http://www.ed.uiuc.edu/EPS/PES-Yearbook/92_docs/Diller.HTM#fn1
Gasana, E., et al. (1999). Rwanda. In Adedeji, A. (ed.), *Comprehending and mastering African conflicts: The search for sustainable peace and good governance* (pp.141–173). New York: Zed Books.
Freire, P. (1998). *Pedagogy of the heart.* New York: The Continuum Publishing Company.
Gergen, K.J. (1995). *Social construction and the transformation of identity politics.* Retrieved on May 5, 2004 from http://www.Swarthmore.ed/SocSci/kgegen1/recent.html.
Gyezaho, B. (2004). The salient social groups in society. *The New Times*, Weekender, August 27–29, p. 8.
Hinchcliff, J.C. (1996). Reconstructing our myths and mindsets for the new millennium. In Slaughter, R.A. (Ed.), *New thinking for a new millennium.* London: Routledge.
Hin, T.S. (1997). *Education for peace: Towards a millennium of well-being.* Paper presented for the working document of the International Conference on Culture of Peace and Governance, Maputo, Mozambique.
KIE (2000). *Teacher development for schools in Rwanda: Resource mobilisation document.* Kigali: Kigali Institute of Education.
———. (2003). *Life skills development program report.* Kigali: Kigali Institute of Education.
———. (2004). *Philosophy of education course assignment.* Groups 19 and 24.
———. (2004). *Philosophy of education course assignment.* Group 15.
———. (2004). *Philosophy of education course assignment.* Group 14.
———. (2004). *Philosophy of education course assignment.* Group 10.
———. (2004). *Philosophy of education course assignment.* Group 22.
———. (2004). *Philosophy of education course assignment.* Group 6.
Linden, I. (1999). The role of non-African NGOs in African conflicts: The case of Rwanda. In Adedeje, A. (Ed.), *Comprehending and mastering African conflicts: The search for sustainable peace and good governance* (pp. 282–296). New York: Zed Books.
Maritim, E.K. and Katia, S.K. (1998). *The project document for UNDP on Kigali Institute of Education.* Kigali: Kigali Institute of Education.

Murenzi, R. (2002, May). *Education policy and vision 2020.* Speech presented at The National Conference on Curriculum in the Service of National Development: What Skills do Our Children Need? Kigali Institute of Science and Technology, Kigali, Rwanda.

Njoroge, G.K. (2002). *The teacher in Rwanda and the challenge of peace building in schools.* Paper presented at the World Teachers' Day Seminar held at Kigali Institute of Education, Rwanda.

———. (2004). *The what and why of holistic learning and living in Rwanda.* Unpublished paper written for the life skills development program of Kigali Institute of Education.

Noddings, N. (1984). *Caring: A feminine approach to ethics and moral education.* Berkeley: University of California Press.

———. (1992). *The challenge to care in schools: An alternative approach to education.* New York: Teachers College Press.

Nussabaum, M.C. (1997). *Cultivating humanity: A classical defense of reform in liberal education.* Cambridge, MA: Harvard University Press.

Republic of Rwanda. (1998). *Study of the education sector in Rwanda, Revised Edition.* Kigali: Ministry of Education.

Rutayisire, J. (2004). *Education for social and political reconstruction: The Rwandan experience from 1994 to 2004.* A paper presented at the BAICE conference, Sheffield University.

Rutayisire, J., et al. (2004). Redefining Rwanda's future: The role of curriculum in social reconstruction. In Tawil, S. and Harley (Eds.), *Education, conflict and social cohesion* (pp. 315–374). Geneva: UNESCO–IBE.

Shapiro, S.B. (2001). *The commonality of the body: Pedagogy and peace culture.* Retrieved June 15, 2004 from http://construct.haifa.ac.il/~cerpe/repapers/sherryshapiro.html.

Shotter, J. (1997). The social construction of our "inner" lives. *Journal of Constructivist Psychology.* Retrieved April 3, 2004 from URL=http://www.massey.ac.nz/~alock//virtual/inner.htm

Woods, E. (2002). *Lecture to Rwanda curriculum conference. Report of the national curriculum conference—Curriculum in the service of national development: What skills do our children need?* Kigali: National Curriculum Development Centre.

# Chapter Sixteen
# Post-Soviet Reconstruction in Ukraine: Education for Social Cohesion

*Tetyana Koshmanova and Gunilla Holm*

### Introduction

Ethnic intolerance is a major obstacle to the development of a truly democratic and inclusive society in any postcommunist country. The numerous ethnic conflicts that have occurred in recent years in various regions of the post-Soviet world—in Russia, Pridnestrovskaya Republic, Moldova, Georgia, Abhasia, Chechnya, and the bloody conflicts in the former Yugoslavia—amply attest to the seriousness of this problem and to the urgent need for teaching interethnic tolerance and acceptance. It is important that education policymakers, university faculty, and teacher candidates understand the potential dangers of interethnic intolerance in order for them to better prepare the next generation of teacher educators, teachers, and students. We suggest academic service learning as a viable approach to further social cohesion via teacher education. Even though the academic service-learning model comes from the United States of America we argue that it can be applied in post-Soviet Ukraine to further accommodate different ethnic groups.

### Analysis of Educational Policies of Ukraine

Since independence (1991), educational and cultural policy in Ukraine has fallen to the "national-democrats," or groups espousing a nationalist ideology. By targeting education for reform, political leaders capitalized on the potential of schools to instill ethnocentric values and articulate norms and traditions of Ukrainian culture. At the first stage of educational reforms their goals were to introduce the spirit of nationalism in order for every person to understand his/her national identity. Nationalist leaders in post-Soviet Ukraine argue that a top priority for the new state must be to protect the; Vyshnevs'ky, 2001). This has been the rationale for launching a "nationalizing" project, an effort to convert a nationalist ideology into an institutionalized national culture, so as to mold a new Ukrainian nation from a heterogeneous population.

However, in the case of Ukraine, this interpretation of the state-formation process is extremely misleading and, 'if pursued to its logical conclusion by the new political elites in charge of the process of state building, is likely to have counterproductive if not catastrophic results' (Batt 1998, p. 57). As Batt (1998) argues, "Ukraine is not and cannot be a "nation-state," not only because it contains a sizable minority of Russians with deep historical roots on the territory, but also because Ukrainians themselves are far from constituting a coherent and unified nation. The most obvious sign of this is the fact that a large proportion of Ukrainians use Russian rather than Ukrainian as their first language" (pp. 57–58). There are three major linguistic groups in Ukraine: Ukrainian-speaking Ukrainians in the west, Russian-speaking Ukrainians in the east, and Russians mainly in the north, east, and south. But it is also important to note that a majority of Ukrainians and Russians are bilingual, so the borders between these two groups are fluid and changeable.

Besides Russians and Ukrainians, there are distinctive groups of Poles, Hungarians, Romanians, Bulgarians, Germans, Greeks, Jews, Checks, Slovaks, Tatars, and Armenians who also have historical roots in Ukraine and compose a considerable part of its population. There are numerous ethnic minorities living in Ukraine, who traditionally were not even considered to be ethnic minorities. In previous years, when the geopolitical map of Europe had a different look, some of these minorities composed a part of the ethnic nucleus of other nations; for example, the Romanians and the Moldavians of Northern Bukovina and South Bessarabia before the beginning of World War II. The same can be said about the Polish people of western Ukraine, and about the Slovaks of the Transcarpathians. During recent years, Ukraine has accepted immigrant groups from Asia and other countries, mainly citizens from former Soviet republics who are leaving regions of conflict and economic instability in search of a better life.

This effort to create a new Ukrainian citizen has fostered the aggressive assimilation of people of other national and cultural identities who have inhabited the country for centuries along with the "pure" Ukrainians (Magocsi, 2002; Verbitskaya, 2003). This assimilationist policy created many challenges to the goal of solidifying peace in the Ukrainian postcommunist setting; it generated bloody religious conflicts between Greek Catholics and orthodox Christians, and created ethnic, political, and social intolerance and instability among multicultural populations. As Tolochko (2001) writes, "There are forces in Ukraine that feed interethnic conflicts. The laws against chauvinism that were adopted in the country are almost not used in the courts, and government agencies that are responsible for controlling the fulfilment of these laws are too tolerant of the chauvinists" (p. 3).

During the Soviet era, the educational system, and in particular the strict controls on historical interpretations played a critical role in the Bolshevik attempt to create *homo sovieticus*. It served as a catalyst for russification and acceptance of an ideal "new kind of human being." The Soviet educational system's highly politicized and Russian-oriented curriculum facilitated communication by conveying shared Soviet cultural values, symbols, and practices among diverse peoples. It was also a key means of indoctrinating a multiethnic population into Soviet ideology and of culturally integrating them to form the "Soviet people." The patterns of thinking and behavior that provided the cultural underpinning of a Soviet identity were taught in school.

In a similar fashion, schools are being called upon to inculcate a Ukrainian identity among the first post-Soviet generation. To foster the development of students' Ukrainian nationality, the educational system was swiftly reformed after independence. The most significant structural changes have been fundamental and sweeping, including, for example, a sharp reduction in the number of schools with a Russian language based curriculum in Eastern Ukraine, and nearly a total closing of such schools in Western Ukraine where Russian as a second language was eliminated in all the schools, and optional classes for secondary students who want to learn Russian were forbidden.

Ethnic intolerance is especially evident in western Ukraine and its major multicultural city of L'viv, where minorities do not feel safe even today. Nationalistic forces are especially powerful here. According to the Russian community leader Elena Bulycheva, during the years since independence, the Russian Cultural Center situated in L'viv is violently attacked by Ukrainian nationalists approximately 2–3 times a year, with the most recent violent attack in June of 2004 (personal communication, July 24, 2004). They have written phrases like "the best Russians are dead Russians," or draw fascist swastikas on the walls on the Russian Cultural Center. According to the leaders of the Russian community movement, Vladimir Provozin (2003) and Alexander Svistunov (1999), during the years following independence a sharp worsening of interethnic relations has taken place in western Ukraine.

Disturbingly, some of the state-sanctioned efforts to impart a collective Ukrainian identity have evolved into a common belief about the uniqueness of Ukrainian national mentality, human traits, and soul, along with an unquestionable superiority over other cultures, especially Russian and Polish. This, in turn, has led to the new "iron curtain" towards Russian, Polish, Jewish, Greek, Romanian, German, Check, Hungarian, Latvian, Tatar, Armenian, and other ethnic minority cultures inhabiting Western Ukraine. Through the socialization process at schools, the cultural minority children acquire knowledge only about Ukrainian culture, language, and history.

The socialization function was always an important consideration in Soviet education, since it was central to the communist ideological control. Now this ideological control is central to the Ukrainian educational program. The notion that Ukrainian universities and secondary schools are responsible for molding the students' personality into "pure Ukrainians" has become a cornerstone of teacher education. (Vyshnevs'ky, 2001). To redefine the Ukrainian nationality, new ideologists count also on preservice education as a main tool to convert the nationalist ideology into an institutionalized national culture and a meaningful mainstream national identity through worshiping nationalistic idols (Fisher, 2002). The same can be said about history textbooks for secondary school students, which teach them "nationalism, ethnic intolerance, and chauvinism towards minorities" (Verbitskaya, 2003, pp. 2–3).

We contend that by changing the contemporary educational paradigm of posttotalitarian Ukraine, to multiculturalism and an inclusive approach to education, educators might diminish the obstacles to the creation of a civic society, based on the postmodernist philosophy of open dialogue among cultures emphasizing democracy, equity, and social justice.

Research shows that the study of Ukrainian nationalism has been often bedeviled with simplistic assertions (Wilson, 1997). Research shows that the main nationalist organization of previous century, the Organization of Ukrainian Nationalists founded in 1929, gained an unsavory reputation as a pro-Nazi and anti-Semitic authoritarian movement (Magocsi, 2002). Observers of contemporary Ukraine have too easily accepted the opposite point of view that all modern-day nationalists are model democrats and civic-minded liberals (Wilson, 1997). The issues of national intolerance may be the main obstacles to democratic reforms in Ukraine on its way to the European Union.

Claudia Fisher (2002), analyzing essays on the history of student–award winners of the School Students' Competition on Ukrainian History, Slidami Istorii, shares similar concerns about Ukrainian ethnocentrism evident in their writings, "Do young people need heroes and models to find one's identity? If so, the responsibility and the task of those who are entrusted with the development of curriculum, textbooks, in-service teacher training and teaching of history should consist of helping young people to find orientation, but avoid giving them heroes and idols strengthening nationalism, racism or totalitarianism" (p. 15).

Even Olha Sukhomlyns'ka (2001), a highly respected Ukrainian educational researcher and official, analyzing dilemmas of Ukrainian education, had to admit,

> Contemporary Ukrainian education and pedagogy are facing a real crisis . . . Having declared a complete change of educational paradigm, pedagogues actually didn't fully elaborate its fundamental standards and concepts . . . In many cases, they just substituted communist with Ukrainian nationalistic ideals . . . For instance, the communist ideology idol, Vladimir Lenin, was substituted with the Ukrainian idol—national leader and writer Taras Shevchenko; the Soviet school's ultimate goal—to develop a communist friendly and monolithic collective, united by common values, was transformed into the goal of shaping a nationalistically minded collectivity out of all Ukrainian citizens. This process can be easily observed in doctoral dissertations. We are experiencing a coming back to old dogmas and stereotypes. (p. 2)

According to Sukhomlyns'ka (2001), 90 percent of all the doctoral dissertations defended in Ukraine during 1991–2001 were devoted to learning history of the Ukrainian education, especially the lives of its personalities, and such a situation did not change much in 2004 either (Sukhomlyns'ka, 2004).

After the extensive use of past Ukrainian education as a theme and its orientation toward glorification of idols, a new research trend can now be seen in connection with the events of the Orange Revolution. Spontaneous democratic movements of people during presidential election in November–December 2004 caused by common dissatisfaction with massive fraud and traditional government policy oriented toward preserving Soviet past symbolized by the unity with Russia clearly showed people's desire toward integration with Europe.

These trends promote the development of a pedagogical research agenda searching for common standards with European teacher education. This process led to Ukraine's formal participation in the Bologna process in May 2005. This new research has prompted a growing interest among educators toward professional preparation of teachers in the light of European educational goals of global citizenship and educational compatibility among teachers. Though the support for nationalism is still strong, the pro-European orientation among Ukrainian teacher education professionals changes the themes of doctoral dissertations defended in 2004–05. Our research shows an increase in topics about tolerance, inclusion, and civic education by approximately 35 percent.

### Research on and Need for Multicultural Education

Teacher education students are currently starting to speak about tolerance, about the necessity to listen to, appreciate, and understand others (Koshmanova, 2002). The need for multicultural education has not been recognized yet among either students or most of the faculty. Students' understanding of tolerance is very elementary in the sense that it is perceived as responding in a passive and patient way to people who may have different points of view.

Since independence and until recently, perspectives focused on tolerance and peace have not been on the agenda of Ukrainian educational research. The theoretical and practical research on peace, justice, and democracy conducted during the Soviet times (with most noticeably the Program for Peace adopted in 1972 by the Twenty-Fourth Congress of the Communist Party of the Soviet Union), along with cultural-historical research on activity theory, learning environment, dialogue of cultures, and education for promoting friendly relations in the classroom (represented by the ideas of Vygotsky [1978], Leontiev [1977], Makarenko [1984], Bakhtin [1979], Bibler [1988] and many others), and research on humanism and democracy in the classroom has been ignored because of the Soviet political propaganda that this research included (Sukhomlyns'ka, 2001). Only recently Ukrainian scholars (Clark and Koshmanova, 2000) have begun to return to this research but with a critical approach.

Educational research conducted by Ukrainian educators and administrators in 2002–04 shows a certain degree of recognition of the ideas of tolerance and multicultural learning (Sukhomlyns'ka, 2004). However, the first serious step in this direction was taken in the middle of the 1990s when the Academy of Pedagogical Sciences of Ukraine started to publish the refereed educational journal *Pedagogy of Tolerance*. Though the number of similar kinds of journals is slowly growing (for example, *Ridna Shkola* [Native School], *Shlyah Osvity* [The Path of Education]), they are still fairly old fashioned with regard to their views of multiculturalism.

Though some educators recognize the importance of the issue of tolerance (Bezkorovayna, 2003; Shvachko, 2000), many of them discuss tolerance and multicultural issues only in terms of theory without considering ways of introducing this concept into the practice of classroom teaching and learning. Shvachko (2000), for example, argues that tolerance "denotes the active participation of a student in accepting the social reality in favor of another person" (p. 7). Another researcher asserts, "Tolerance is something different than passiveness and enduring violence, pain, or negative influence... Tolerance implies respect for equity, recognition by others, refusal to be subjected to domination or violence" (Pastushenko, 1999, pp. 28–29).

However, there has been no Ukrainian research conducted on the issues related to student ethnic intolerance or strategies to overcome prospective teachers' ethnic stereotypes. Even though there is a recognition among researchers that "ethnic intolerance, xenophobia, and ethnocentrism . . . exist" (Verbitskaya, 2003, p. 2), others argue that the understanding of "the importance of multicultural education still exists mainly at the declarative level" (Provozin, 2003, p. 3).

Despite the majority of teacher educators' orientation toward nationalism and the domination of educational stereotypes in teacher educators' thinking, there has always been a tendency toward a democratic education in the light of the best traditions of the cultural-historical activity approach in teacher education that started in the Soviet Times (Bakhtin, 1979; Bibler, 1988; Vygotsky, 1991). This tendency continues to influence many contemporary Ukrainian teacher educators (Koshmanova, 2005). Today liberally minded teacher educators are supported by people who supported the Orange Revolution. However, as in any teacher education institution there will be teacher educators in favor of a monocultural approach as well as a multicultural approach to the preparation of teachers.

## The Development of a Multicultural Perspective among Monocultural Teacher Education Students

We argue that three approaches would strengthen Ukrainian teacher education with regard to promoting tolerance. First, a democratic classroom is essential if we want to build a democratic society. Students learn democratic principles by practicing them. Second, students need to learn to care about those who are different. Caring, however, is not enough, but needs to be combined with critical thinking in order for students to become activists working for acceptance of cultural diversity. Third, we argue that academic service learning is a way for students to confront on their own identities, stereotypes, and prejudices.

### A Democratic Classroom

In a democratic classroom the instructor models both the intellectual and moral virtues that preservice teachers will need to learn to model and exercise themselves as practicing teachers in a democratic, culturally diverse society. The intellectual virtues that would promote open communication include encouraging questioning, listening carefully and respectfully, and analyzing statements in order to better understand others. Students need to learn that it is safe to take a stand and to evaluate positions. Teachers need to model critical thinking and expressing their views. Thanks to the Orange Revolution, European integration and Bologna process, it is becoming easier for teachers to create the kind of classroom atmosphere where students and teachers feel safe to express themselves.

Likewise, the instructors need to model giving constructive criticism. The moral virtues include among others, modeling tolerance of and respect for others and their views as well as caring about others' points of view. However, it is important to note that the instructor is not an equal of the students. The instructor bears the responsibility for developing a democratic classroom, not a relativistic classroom where all views are regarded as equal. The goal for teacher education is to contribute to building a democratic and peaceful society, not an ethnocentric and nationalistic society.

The instructor, together with the students, needs to strive to build a classroom community where the participants learn to critically examine both the written and lived curriculum for intolerance, ethnocentrism, and bigotry. In other words, students need also to examine critically their own as well as their fellow students' views. This can be a difficult and painful experience to acknowledge one's own stereotypes and ethnocentric views. It is helpful for students in this process to understand their own locally grounded experiences in the light of a larger, more global perspective.

In order for multicultural education to flourish, classrooms have to be democratic both at the university and the grade school levels. Teacher education students need to understand their own roles as change agents in a democratic society. Teachers are key professionals in socializing new generations of young people toward taking a stand for a democratic, culturally diverse society as opposed to living passively in a nationalistic, ethnocentric society. Future teachers need to learn to be more in charge of both the formal and the hidden curriculum in order to counteract other more nationalistic influences.

## Critical Thinking and Caring

The traditional way of teaching in the Ukraine may be considered to be what Freire called "the banking model" (Freire, 1970). It is a model that does not challenge students to think critically or to examine their own views. Learning according to this model is a fairly passive activity of receiving a body of knowledge from the expert teacher. Some changes to this model are slowly appearing in Ukrainian teacher education. For example, many (Karas, 2005; Hrynkevych and Tsyura, 2005; Ravchyna, 2004) now emphasize the necessity for teachers to organize student learning based on student activities for achieving socially meaningful goals.

Applying this concept to teaching tolerance and multiculturalism it is useful to base it on the ideas of Vygotsky (1978) and Makarenko (1984) about communication as a cooperative activity that recently came into the international prominence. The constructed activities should be grounded in the students' own interests. To achieve the goals of developing students' levels of tolerance and multicultural understanding an instructor could use the system of the two motives: one of them is a concrete motive, which is personally meaningful for students. This motive, based on students' immediate needs such as recognition from others, interesting communication with friends, a good grade for the assignment, and so on becomes the primary goal for doing a certain activity. The second motive of socializing with others is more meaningful although it is not perceived as a meaningful motive in the beginning. However, such a motive can become an important goal, if the student has a positive personal experience by participating in the activity (Ravchyna, 2004).

In our seminar where teacher education students were assigned to do fieldwork in a culturally different community, we found in their reflection journals that students were indifferent or opposed to the ideas of multiculturalism in the beginning and did not want to get involved in a cultural activity in an ethnic minority community. But they wanted to fulfill the assignment given by the instructor, their primary motive being the grade for the course, as well as spending more informal time with their groups mates, who were also assigned to be in the ethnic community. Through enjoyable communication with friends, work on an assignment grade, personal experience in the activity (visiting the community several times), and communication with minority children, students gradually began to realize that they liked this activity. Their biases and stereotypes toward this ethnic minority became smaller, or even diminished and disappeared.

In other words, the secondary motive became as powerful as the primary motive thanks to the pleasurable activity where students let their guard down and at least temporarily forgot their ethnic biases and stereotypes. As witness to this are the following excerpts from the journals students wrote:

> Only now I have understood how pleasant helping people could be, especially helping children who need caring... I cannot forget those happy and grateful eyes, which looked devotedly at me. Exactly this look helped me understand and realize the usefulness of the work I did. But at the beginning, it was just completing the assignment, it was important for me to do the task my instructor needed. Now my attitude towards this has changed. I want to continue doing something new and useful.

> I went to this community just to get an excellent grade. To be honest, I didn't know why it was needed. You know, I believed that it was not necessary to go to the minority community to fulfill this assignment. Indeed, maybe it is not pleasant to me. I could successfully complete the task in another place. But somehow, in a strange way, my communication with friends began gradually to engage me in the process of working with children. At one moment, those kids seemed to be not that bad... And what is special about their being different? They are the same as others: they want love, caring, attention. And at this moment, I thought that I could also be perceived by someone of a different ethnicity as not really friendly.

> I even didn't know that there are some ethnic differences among the Ukrainians, some different customs and traditions, and that some of the "Ukrainians" are not Ukrainians at all. Working in schools for minorities, I got convinced that students better open to each other while telling about their culture and lives.

These excerpts from students who participated in fieldwork in a culturally different community indicate that they became more aware of others as well as started to care somewhat for others. Caring is an important characteristic for teachers especially learning to care about marginalized students. However, in a society that is driven by ethnocentric public policies caring about those who are excluded is not enough. Teacher education students also need to learn to think critically about the societal structures and policies that exclude and marginalize the children and youth of culturally different backgrounds.

## Academic Service Learning

In this section we argue that academic service learning (ASL) is a promising way to promote the understanding and acceptance of culturally different groups among teacher education students. ASL has been used in the United States of America, for example, to better prepare white middle class, and mostly female teacher candidates from suburban and rural backgrounds to teach inner-city students from a different racial and ethnic background. Although ASL has been mostly used in the U.S. context, we think it would be an important approach for Ukrainian teacher education to promote among its teacher education students the acceptance of ethnically different groups. The ASL has its roots in Dewey's ideas about experiential learning but today also in Freire's critical pedagogy (Deans, 1999). From the beginning ASL has been an activist approach. In other words, accomplishing change is central to the mission of ASL (Stanton, Giles, and Cruz, 1999). For many universities ALS is also a way to promote citizenship education and participation (Bringle and Hatcher, 1996; Bringle, Games, and Malloy, 1999; Rhoads, 1998). A survey conducted in the United States of America in 1998 "found that more than 225 of the approximately 1,325 teacher education programs in the nation offer service-learning experiences" and over half the number of high schools also offered or required service-learning experiences (Anderson, 2000, p. 1). These figures have increased substantially over the years due to the many grants that have been made available for developing ASL courses.

The goal of ASL courses is to tie the academic content of a class to issues of importance in a community in order to make the academic content more relevant for students. The service the students engage in should be based on a need in the community and should be beneficial for both the community members and the students. Hence, the service-learning activities have to be developed in collaboration between the university and the community. In other words, the university students' work is not simply an exercise on which they are graded but an actual authentic issue that is important to people in the community. In preparing students to work with community members it is necessary to deconstruct some common myths among university students (Langseth, 2000). These myths center on the idea that the university culture is superior to the local community cultures and that faculty and student knowledge is superior to community members' knowledge. Furthermore, university faculty are often considered wiser than community members by the students. Instead university and community members have to treat each other as equals. This is also a healthy thing for prospective teachers to learn with regard to their future interactions with parents. Community members know their communities and parents know their children in depth. Overall, students need to learn to focus on community assets and not just to see the deficits. The thought that they will learn from the community is often new to the students but the experience contributes to an increased respect for the community among the students (Howard, 2001). For these reasons, ASL is considered a major vehicle for teaching civic education to young people (Seals).

Interacting with community members helps the university students understand how others see them, which in turn helps them realize how they view others in terms of stereotypes and prejudice. They come to understand that how others see them influences the work they can accomplish in the community and that they have to learn to interact in culturally sensitive ways (Rice and Pollock, 2000). Students need to be challenged to confront their own stereotypes and prejudices. An important part of this work is the ongoing reflection upon their work. This can be done through classroom discussion, reflective journals or analytic papers. Not reflecting or being challenged about their stereotypes can actually lead to ASL confirming existing stereotypes and prejudices among students.

From our experience with ASL in one of the largest teacher education programs in the United States of America we have found that it has a major impact on teacher education students (see Holm and Farber, 2005). We have incorporated ASL into a teacher education course on social justice in education. Teacher education students worked as mentors in both an alternative high school for students who had failed in the regular high schools and in an elementary school. In both schools most of the students were poor and the majority were either African American or Latinos. Most of the students were not doing well academically and very many had difficult home circumstances due to a large extent to the poverty of the family. Hence, the teacher education students were engaged as mentors or adult friends to these students. They helped with academic work at times or simply listened to the students when they needed to talk. They also introduced the students to the idea of going to college since many of the students came from homes where no one had had any contact with the university. In most cases the students had never been to the university' campus or to any cultural or sports events organized by the university even though the university is located only a few kilometers from the schools. The teacher education students on the other hand were almost exclusively white and middle class. All of them had been quite successful academically. Hence, the students were placed by the school personnel as mentors to students who were very different from themselves both culturally and academically. This posed a challenge to them especially since many of the students did not speak standard English which was a new experience for the university students.

For many of the university students it was the first time they were in regular contact and built a relationship with a person of color or a person living in poverty (Boyle-Baise and Kilbane, 2000). It was often a challenge for the university students but once they had worked with one student they at least knew what some of the issues were going to be when they later would have their own classrooms with 25 students from all kinds of backgrounds and cultures. The university students also learned that diversity means more than ethnicity and race. Social class and gender constituted chasms that do exist in all schools that they too had to at times bridge. The students wrote weekly electronic reflection journals and participated in an electronic discussion about the issues they encountered with regard to the students and their families, the school structure and curriculum. In both schools the curriculum resembles what Haberman (1991) calls the pedagogy of poverty with worksheets to keep the students busy and under control. Learning is often not the focus of such a curriculum. These kinds of issues made the university students question their own privileged schooling that they, for the first time, understood to be privileged. The university students had always had supplies, good teachers and facilities, up-to-date textbooks, extracurricular programs, and fundraising help from their parents and communities. This was an essential feature of the ASL experience because in order for the students to not blame individual students and their families for the difficulties they have, they themselves need to understand that many of the problems are structural. However, one of the most difficult aspects for the university students was not to blame the families for the students' academic failures and other difficulties. The university students struggled with understanding that in most cases the families did the best they could. Sometimes the problems came from the fact that the parents worked several jobs in order to support the family and had little time for their children. Seeing clearly the assets in the families and communities or even in the schools did not come easily, but ASL did contribute to their doing so.

Not much is known about how ASL impacts students' understanding of academic learning, but it is very clear that it has a substantial impact on students' attitudes and acceptance of cultural diversity (see also Root and Furco, 2001). Astin, Sax, and Avalos (1999) also found that it leads to more community engagement. In other words, students participating in ASL will be more active citizens, something that is crucial for an emerging democratic society like Ukraine.

## Conclusions

The political, cultural, and educational policies in post-Soviet Ukraine are taking the country in an ethnocentric and intolerant direction. This is of particular concern since Ukraine, as well as most of the other countries in the former Soviet bloc, does not have an ethnically, linguistically, or religiously homogeneous population. The various minorities have no space in the current Ukrainian society.

We argue that changes toward being a country that is accepting and appreciative of cultural diversity could be built from a grassroots level through the educational institutions. Teacher educators and consequently teachers themselves can actively contribute to building a democratic and tolerant post-Soviet Ukraine. By practicing democracy in their classrooms young people and children will learn what democracy is. Furthermore, engaging in academic service learning contributes to students' civic engagement. Interactive learning strategies including simulations, cooperative learning, and constructive criticism require an atmosphere of trust and would forge a socially cohesive learning community, which teacher education students could later build again in their own classrooms. Teacher educators and their students can also further the development of a multicultural society by confronting and reflecting on their own positions, privileges, stereotypes, and attitudes in contemporary Ukrainian society. Teachers who care about culturally different students as well as critically think about and work against the structural barriers creating unfair conditions for those who are not considered "hundred percent Ukrainian," are the building blocks for a culturally diverse democracy.

## References

Anderson, J. (2000). Service-learning and preservice teacher education. *Learning in deed issue paper*. Denver, CO: Education Commission of the States' Initiative, Compact for Learning and Citizenship.
Astin, A.W., Sax, L., and Avalos, J. (1999). Long term effects of volunteerism during undergraduate years. *Review of Higher Education*, 22 (2): 187–202.
Bakhtin, M. (1979). K metodologii gumanitarnyh nauk [Towards the methodology of humanitarian sciences]. *Estetika slovesnogo tvorchestva* (pp. 361–373). Moscow: Iskustvo.
Batt, J. (1998). National identity and regionalism. Introduction. In T. Kuzio (Ed.), *Contemporary Ukraine: Dynamics of post-Soviet transformation* (pp. 57–59). New York: M.E. Sharpe.
Bezkorovayna, O.V. (2003). Z pedahohikoyu tolerantnosti u trete tysyacholittya" [With pedagogy of tolerance into the third millennium]. *Pedagogy of Tolerance*, 3 (25): 4–12.
Bibler, V.S. (1988). Shkola dialoga kultur [The Dialogue of cultures school]. *Pedagogika*, 11: 29–34.
Boyle-Baise, M. and Kilbane, J. (2000). What really happens? A look inside service-learning for multicultural teacher education. *Michigan Journal of Community Service Learning*, 7: 54–64.
Bringle, R.G., Games, R. and Malloy, E.S. (1999). *Colleges and universities as citizens*. Needham Heights, MA: Allyn & Bacon.

Bringle, R.G. and Hatcher, J.A. (1996). Implementing service learning in higher education. *Journal of Higher Education*, 67 (2): 221–239.
Clark, C. and Koshmanova, T. (Eds.). (2000). *Culture and community in learning to teach.* L'viv: University of L'viv Publishers.
Deans, T. (1999). Service-learning in two keys: Paulo Freire's critical pedagogy in relation to John Dewey's pragmatism. *Michigan Journal of Service Learning*, 6: 15–29.
Fisher, C. (2002). Can one write history without a heroic pathos? Notes suggested by checking the best student competition papers on the history of Ukraine. *Doba in Civic Education*, 2: 7–15.
Freire, P. (1970). *Pedagogy of the oppressed.* New York: Herder and Herder.
Haberman, M. (1991). The pedagogy of poverty versus good teaching. *Phi Delta Kappan*, 73: 290–294.
Holm, G. and Farber, P. (2005). Cultural competence: College students learn through academic service-learning. In G. Holm and H. Helve (Eds.), *Contemporary youth research: Local expressions and global connections* (pp. 121–132). Hampshire, UK: Ashgate.
Howard, J. (2001). *Service-learning course design workbook.* Ann Arbor: University of Michigan OCSL Press.
Karas, A. (2005, July). *Hard road to civil society and common Europe: The case of Ukraine.* Paper presented at the ICCEES VII World Congress Europe—Our Common Home? Berlin: Humboldt University.
Hrynkevych, T. and Tsyura, S. (2005). Educational interaction in specific multicultural environment. *The University of L'viv Referred Scientific Journal: Pedagogical Series*, 19: 167–175.
Koshmanova, T.S. (2002). Development of teacher education in the United States: 1960–2002. Doctoral Dissertation, Kyiv: Instytut Pedagogiky i Psyhologiyi Profesiynoyi Osvity.
———. (Ed.). (2005). *Pedagogy for democratic citizenship.* L'viv, Ukraine: Publishing House of Ivan Franko National University.
Langseth, M. (2000). Maximizing impact, minimizing harm: Why service-learning must more fully integrate multicultural education. In C.R. O'Grady (Ed.), *Integrating Service-Learning and Multicultural Education in Colleges and Universities* (pp. 247–262). Mahwah, NJ: Erlbaum.
Leontiev, A.N. (1977). *Deyatel'nost'. Soznanie. Lichnost'* [*Activity. Consciousness. Personality*]. Moscow: Izdatel'stvo politicheskoy literatury.
Makarenko, A.S. (1984). *O moyom opyte* [*About my experience*], vol. 4 (pp. 248–267). Moscow: Pedagogika.
Magocsi, P.R. (2002). *The roots of Ukrainian nationalism.* Toronto, London, Buffalo: University of Toronto Press.
Pastushenko, N.P. (1999). "Diahnostuvanny navchennosti uchniv z ukrains'koi literatury yak shlyah do utverdzhennya pedagogiky tolerantnosti: teoria i practyka" [Diagnosting student learning in the Ukrainian literature as a way of celebrating pedagogy of tolerance: Theory and Practice]. *Ukrains'ka mova i literatura v shkoli* [*Ukrainian language and literature in school*], 2: 28–33.
Pidlasyi, I. (2002). *Pedahohika.* Lviv: Svit.
Provozin, V. (2003). *Nash Russkiy Dom* [*Our Russian House*]. L'vov: AHIL.
Ravchyna, T. (2004). Innovative approaches to teaching and learning in contemporary Ukrainian schools. In Ryschard Kucha (Ed.), *European integration through education* (pp. 503–511). Lublin, Poland: Maria Curie-Skłodowska University.
Rhoads, R.A. (1998). In the service of citizenship. *Journal of Higher Education*, 69 (3): 277–297.
Rice, K. and Pollock, S. (2000). Developing a critical pedagogy of service-learning. In C. O'Grady (Ed.), *Integrating service learning and multicultural education in colleges and universities* (pp. 115–134) Mahwah, NJ: Lawrence Erlbaum.

Root, S. and Furco, A. (2001). A review of research on service-learning in preservice teacher education. In J.B. Anderson, Swick, K.J., and Yff, J. (Eds.), *Service-learning in teacher education.* Washington: AACTE Publications.

Seals, G. (Ed.). (n.d). *Service-learning in teacher education: A handbook.* The National Service Learning in Teacher Education Partnership.

Shvachko, O. (2000). Tolerance as social value. *Practical Psychology,* 5: 7–9.

Stanton, T. K., Giles Jr., D.G., and Cruz, N.I. (1999). *Service-learning. A movement's pioneers reflect on its origins, practice, and future.* San Francisco: Jossey-Bass.

Sukhomlyns'ka, O.V. (2001). To the issues of etymology of pedagogical knowledge. *The Path of Education,* 1: 2–7.

———. (2004). Peculiarities of the contemporary transformational changes. *The Path of Education,* 2: 2–6.

Svistunov, A. (1999). *Vo Ves' Rost [In Full Height].* L'vov: Civilizacia Publishing House.

Tolochko, P. (2001). This is our common land. Retrieved July 26, 2004 from http://www.jewukr.org/observer/jo1518/p0601r.html.

Verbitskaya, Y. (2003). How to create conscious Ukrainian from the fifth-graders? Retrieved July 26, 2004 from http://zaistinu.ru/ukraine/press/antihistory.shtml.

Vygotsky, L.S. (1978). *Mind in society.* Cambridge, MA: Harvard University Press.

———. (1991) *Pedagogicheskaya psikhologia* [Educational Psychology], (Ed.) V. Davydov. Moscow: Pedagogika.

Vyshnevs'ky, O.I. (2001). *Teoretychni osnovy pedagogiky: Kurs lektsiy.* Drohobych: Bidrodzhennya.

Wilson, A. (1997). *Ukrainian nationalism in the 1990s. A minority faith.* Cambridge, Great Britain: Cambridge University Press.

# Chapter Seventeen
# Teacher Preparation for Peace-Building in United States of America and Northern Ireland

*Candice C. Carter*

Teacher preparation for peace-building is comprehensive, especially in contexts of cross-cultural conflicts stemming from ignorance, intolerance, a history of inequitable rights, and a legacy of violence. Dispositions, knowledge, and skills for developing peace comprise the foundation for fairly educating all students in regions with diverse populations and contexts where discrimination and structural violence have been evident. In societies with intercultural conflicts and memories of violence, responsive teacher education promotes critical consciousness of instruction, including analysis of hidden as well as explicit curricula. Different expectations of, and support for, students with diverse identities are examples of hidden curricula. Fostering cognitive flexibility, self-assessment, and reflection that evaluates situations for peace development and individual responsibility are components of teacher preparation in regions with a collective memory of violence. Examples of dispositions developed for proactive peace education include agency in threats to peace and responsibility for addressing the needs of all students. Corresponding skills result from knowledge of and experience with appropriate responses to difficult issues. Knowledge about sources of conflicts as well as methods for constructively resolving them are two other components of teacher training for peace development in territories occupied by social-isms.

Teacher preparation for peace building addresses multiple conflict antecedents such as ethnocentrism, sectarianism, classism, racism, sexism, handicapism, and homophobia. Recognition of the effects of bigotry throughout their history and in their increasingly diverse societies has been a catalyst for educational responses in Northern Ireland and Southern United States of America. In both regions, physical, psychological, and structural violence are ongoing. The review of goals and strategies in peace-oriented teacher preparation precedes a discussion of research on the enactment and outcomes of such training.

## Philosophical and Theoretical Foundations

Multiple theories with perspectives and interests comprise the foundation of peace-oriented education. The purpose of changing structures in a society to reduce conflict is evident in social reconstructionism and critical theory whereas adjusting to and managing conflicts in society are supported by the goals of pragmatism and existentialism (Ozmon and Craver, 2003). With a reconstructionist perspective, education is viewed as a tool for reshaping a society that is ridden with structural violence (Brameld, 1956). For accomplishing a reconstructionist objective, educators reveal impediments to social harmony that need removal (Hudak and Kihn, 2001). For example, identifying and relinquishing one's privilege is reconstructive work for equity that supports social cohesion (McIntosh, 1989). George Counts (1932) suggested that educators forfeit their roles as reproducers of a society's social structure to work as social reformers with a transformative, versus indoctrinating, pedagogy. One structure that peace-oriented education addresses within and beyond school is the entitlement hierarchy of people with different social-class characteristics (Mehan et al., 1996). Another structural approach to peace-focused schooling is detracking, whereby students who have previously been sorted into different levels of curriculum and instruction are integrated in courses with high-level and supportive instruction. Detracking provides more equitable opportunities for social as well as cognitive development and shares the power held by dominators with others who have been objects of structural violence through disempowerment.

Much like reconstructionism that identifies antecedents of structural violence, critical theorists examine power relationships in society with questions such as "Who has the power? What knowledge is transmitted? How is knowledge transmitted? and What are the social divisions?" (Apple, 1990; Finn, 1999). Examination of implied lessons evident in teacher and student interaction, their physical environment, instructional techniques, and curricula used aids with conflict analysis (McLaren, 2003). For example, lack of a family's cultural capital that facilitates their involvement in school decisions about their children is recognizable through analysis of interaction patterns (Bourdieu, 1977; Lareau and Horvat, 1989). Restructuring schools for broad-based parent involvement advances social cohesion in a community when families of different backgrounds work together (Eisler and Miller, 2004). The integrated schools of Northern Ireland that bring together families across sectarian boundaries evidence such restructuring (McGlynn, this volume). Conflict resolution programs in schools that successfully empower students of all backgrounds for solving interpersonal problems is a restructure of approaches that typically put administrators and teachers at the forefront of managing students' relations at school (Carter, 2002a; School Mediation Associates, 2005).

A pedagogical emphasis on interaction experience for learning about and solving community problems characterizes pragmatism in education. John Dewey (1916) advanced the concept of instrumentalism in education whereby students learn ideas in their community and apply them. Learning through the examination of problems and taking social action to address them supports development of prosocial citizenship (Boulding, 1988; Morrison, 2005). Examples of pragmatism include cross-community activities that build bridges between divided populations, student care-taking of people and environments as well as students' international peace projects. While pragmatism promotes relevant knowledge and skill formation as well as initiative for problem solving in local to global contexts, existentialism facilitates self-assessment for conflict resolution.

Existentialist learning promotes examination of patterns in environments. The "wide-awakeness" that Maxine Greene (1973) recommends that educators bring about for their students calls their attention to the circumstances which have been shaping their existence. Students' heightened consciousness and reflective thinking expand their perspectives of problems, which facilitates their understanding and constructive response to them (Carter, 2004c). Broadened perspectives of how people experience life can help students actively listen to how others see and feel. The door of social cohesion opens wider when those having different positions in a conflict listen to each other (Lantieri and Patti, 1996). Peace education promotes active listening and compassionate communication in which disputants focus on each other's needs more than the characteristics that divide them (Rosenberg, 2000).

Recent practices of peace-oriented education evidence the philosophical and theoretical underpinnings of pedagogy that is focused on transforming individuals and structures in their conflict-ridden societies (Morrison and Harris, 2003; Salomon and Nevo, 2002). This chapter describes regions in two countries that have histories of intergroup violence within their borders and involvements in international wars.

### Overt and Subliminal Violence in Southern United States of America and Northern Ireland

Violence in these two contexts has ranged from unintentional harm of others through unconscious discrimination to discrete or blatant verbal and physical attacks (McKittrick, 2002). Physical, psychological, and structural violence continues in Southern United States of America and Northern Ireland, although armed conflict is not currently widespread within their national boundaries. Their national governments wage wars abroad. These conditions along with the prevalence of their media entertainment that usurps peace illustrates normalization of violence.

The antecedents of social conflicts in Southern United States of America and Northern Ireland are evident in their similar histories with political and economic disparity of groups that endured official discrimination for centuries (Byrne, 1997; Hainsworth, 1998; Ortiz, 2005). Occupation and colonization of their homelands forced indigenous peoples throughout United States of America and the Irish to struggle for fair treatment including acquisition of equal social, economic, and judicial rights in their homelands. Many other dominated groups in United States of America, especially former slaves whose physical characteristics are different from those of the dominant group, have endured and struggled to overcome pervasive discrimination. Injustice based on perceptions of race, religion, culture, social class, gender, sexual orientation, and ableness due to physical or cognitive differences continue despite legislative changes to enforce fair treatment and equal representation in at least governmental organizations (Hacker, 1992; Takaki, 1993). While overt violence has characterized struggles against dominance and discrimination in Southern United States of America and Northern Ireland, most notably during the civil rights era in addition to the Civil War in United States of America and the Troubles in Northern Ireland, it currently is not widespread within their boundaries. In Civil War and the Troubles, physical violence was widespread as dominated peoples and their supporters responded to pervasive injustice in their societies. Nevertheless, segregation continues in both regions while the possibility of physical attack is felt on all sides of sectarian, racial, and ethnic boundaries. In Northern Ireland, collective memories of violence experienced during battles between groups in the past half century keep clear in the minds of many the losses they suffered and the dangers that loom in the undercurrents of interaction between historically opposed groups (Kapur, 2004). Evident tensions associated with intergroup relations in Southern United States of America and Northern Ireland are felt by the youth in those societies who learn from their elders and observations in their communities to maintain relationship boundaries. While many learn to cooperate in public and some private locations where they interact, crossing the established social boundaries to form enduring friendships and familial relationships has not become common in either region.

The two contexts described here are at best superficially polite societies in which violence most commonly underlies veneers of cooperation in public. However, scapegoating against diverse immigrant groups has been increasing in Northern Ireland while identifying certain cultures as possible "terrorists" continues in United States of America and the United Kingdom (UK). In both nations, increasing costs of wars abroad have exacerbated intergroup tensions and increased animosity toward cultural minorities.

These patterns of unrest and inter-group conflict are historically evident throughout the United States of America and UK, as well as in other nations where peoples' -isms underlie violence. Hence, educational responses include goals for developing students' prosocial dispositions, skills, and knowledge that support understanding, acceptance of differences, and intergroup cohesion.

## Standards-Directed Education

Influencing teacher training in United States of America is a plethora of standards and guidelines that educators must reference in their lesson plans. Consequently, educational practice, including teacher training, has become more uniform from accountability to official guidelines for instruction and corresponding curricula. Accordingly, it is very limiting when teachers only have or use guidelines that lack the breadth of peace-building recommendations. Teachers who are focused on peace building have had to seek guidelines of multiple organizations, beyond government standards and modules, to reference prescriptions for their prosocial practice.

Recommendations for peace education are found in two types of sources; organizations which maintain that focus (The Anti-Defamation League, 2006; The Collaborative for Academic, Social, and Emotional Learning, 2006; Educators for Social Responsibility, 2006; Transcend, 2006) and others that include societal-cohesion strands in their broader objectives for education (Council for Education in World Citizenship, 2006; National Council for the Social Studies, 2006) Variably addressed across sets of standards for education are recommendations that encourage development of self-knowledge, human diversity, analysis of problems, experience with solving

**Table 17.1** NAME Criteria for Evaluating Curriculum Standards

| *Key Concerns* |
| --- |
| Inclusiveness |
| Diverse perspectives |
| Accommodating alternative epistemologies and social construction of knowledge |
| Self-knowledge |
| Social justice |

problems, and reflection on methods of problem solving. Table 17.1 displays guidelines drafted by the National Association for Multicultural Education (NAME) to use in evaluation of any set of standards that prescribe practice. These standards by NAME for other standards that are influencing practice prescribe more than human diversity education, which is a common recommendation.

NAME's social-justice standard clearly prescribes teaching about interpersonal relations and societal structures that are antecedents of violence. It also recommends training for experience with action taken in response to -isms.

> Promote social action, creating an engaged, active, and responsible citizenry committed to eradicating bigotry and to developing a fair, just, democratic society responsive to the needs of all our people regardless of race, class, gender, age, sexual orientation, physical appearance, ability or disability, national origin, ethnicity, religious belief or lack thereof. (National Association for Multicultural Education, 2006, p. 3)

Unfortunately, most education standards have few guidelines for peace-oriented education. The perseverance that the Anti-Defamation League recommends in training must be practiced by educators for teaching how to effectively respond to -isms (Anti-Defamation League, 2006). Teachers equipped with philosophies that support peace education as well as knowledge of how to include student action in the study of social problems are prepared to persevere in standards-focused contexts. Crucial in teacher education is careful analysis of standards that shape their practices.

## Teacher Preparation

Foci of teacher preparation for social education includes, disposition, knowledge, and skill development. First, teacher-candidates learn values that may be different from those that they have in the beginning of their training programs. For example, intolerance of -isms that undermine equity in education and other aspects of their society is one goal for disposition development (Banks, 1997). Second, knowledge of pedagogy and knowledge of social history and current social contexts are foundations of teacher education (Freire, 1998). Pedagogical knowledge includes methods of teaching that incorporate the rich diversity students bring to their schools as well as the multiculturalism of their local and global communities (Grant and Sleeter, 1998).

Cultural knowledge of beliefs and practices in the community where they teach helps teacher candidates understand the needs of diverse populations with whom they will interact during their career. Teachers-in-training also learn the importance of keeping informed about social concerns in their communities along with being a role model as a proactive citizen (Fulton and Gallagher, 1996). Third, teacher preparation programs assess the skills of their candidates including critical thinking and problem solving for the improvement of social interactions and structures that interfere with peace (National Council for Accreditation of Teacher Education, 2006; Smylie et al., 1999). Skills that need inclusion in standards guidelines for teacher education include maintaining a vision of improved human conditions and relations as well as communicating constructively about impediments to actualizing their vision of peace through education (Reardon, 2001). For example, a needed standard could state that successful teachers evidence their vision of equal opportunities for all students and effectively communicate about conditions that obstruct justice. Another skill of peace-focused instruction is community building in socially divided contexts (Sergiovanni, 1994). Building a community of mutual understanding and trust for teachers in a divided society is crucial for transforming their relationships and modeling peace processes. Unless assisted with changing, teachers are prone to maintaining existing relations and reproducing social conditions through a continuation of thought patterns and corresponding behaviors that they learned in their communities.

Preparing teachers for their own practice of peace-oriented education entails learning about oneself as an individual and as a member of a community where injustice and other roots of violence exist (Hollins, 1997). Examination of one's identities and self-assessment of one's own attitudes, knowledge, and skills that support or undermine peace are initial tasks of peace development. Teacher candidates expand their knowledge of human diversity while they are taught to respectfully recognize and accommodate it where differences of beliefs, values, and habits can cause conflicts. They are also instructed as role models to recognize and manage their thought patterns, emotions, and communication, especially in contexts of conflicts or cultural differences. Such responsibility is not easy for teachers-in-training who are just beginning self-work. Their preparation entails community-based, in addition to classroom-based, practice with self-modulation. Beyond expressing a desire for peace development, personal satisfaction and collective experience with its enactment in teacher education best portend for it future practice. In her discussion of antecedents of voluntary peace building, Noddings (2005) notes that "Caring for, about, and with others can add to our own happiness" (p. 8).

Successful experiences with the educational practices mentioned are assessed throughout teacher preparation programs that identify candidates who evidence a need for more training, such as an inability to transform their mental and physical habits that could perpetuate -isms and other antecedents of community conflicts. Research on teacher training informs the field of its challenges in developing or transforming dispositions, knowledge, and skills.

## Enactment

Between 1999 and 2004 qualitative research on teacher preparation in Northern Ireland and Southern United States of America produced data from interviews, observations, and analysis of writings produced by teachers-in-training, their instructors, as well as by other researchers on this topic. In all sites of data collection, the researcher inquired about practices that were identified by the participants as preparation for peace building. Examination of teacher training programs and childhood education by graduates of those programs rendered emergent data that revealed categories and dimensions of cultural desegregation and peace-oriented instruction. In a Florida university, participant observations occurred with the researcher's practice of teacher education.

## Cross-Cultural Integration

Approaches used in United States of America and Northern Ireland for teacher education are similar, yet different, due to the contexts of each society and their program models. In universities of Northern Ireland and Southern United States of America cross-cultural integration across levels of education varies, mainly as a result of different types of program offerings. For example, the two teacher preparation colleges at one university in Belfast differently attract Catholic and Protestant students whereas the University of Ulster campus integrates students in one preparation program. Segregation in United States of America occurs through differences between public and private universities. Some private universities in United States of America attract mainly African American populations while other private and public universities enroll students of diverse cultural backgrounds. Consequently, integration of teacher candidates who do not cross cultural boundaries by studying together in the same preparation programs presents a challenge for teacher education in universities that lack cultural diversity. Yet, universities with cross-cultural enrollments faced challenges in building more than superficial bridges between historically divided groups. Nevertheless, separated and integrated programs have the same official goals and standards for teacher preparation.

Cross-cultural integration of students and instruction for enhancing multicultural knowledge and skills were mandated for teacher preparation in Southern United States of America, such as a statutory course in Florida titled "Teaching Diverse Populations." The initial program provided across Northern Ireland that was designed for student understanding of and contact with other cultures was Education for Mutual Understanding (EMU). However, teacher preparation for use of EMU was implicit in Northern Ireland. A common approach to implementation of EMU's prosocial themes in childhood and teacher education was integration of divided populations through social contact. Such integration is often the first sustained contact across their society's sectarian boundaries for students who are not learning with culturally diverse peers. Like the teachers in the children's schools, teacher educators in the universities had much flexibility in the use of EMU. At integrated and segregated universities, the author was told by all of the teacher educators she interviewed that use of EMU was a challenge for them to find a good fit in their instructional programs. In addition to time pressures, the challenge may have been partially rooted in a lack of sustained integration experience by some teacher educators, or psychological scars from their losses during the Troubles. The flexibility of EMU implementation presented another challenge in that it left open to interpretation how cross-cultural contact and instruction could be done. Such flexibility was designed for development of best-fit models in the different teacher preparation programs of Northern Ireland. That fit mainly resulted with brief interaction across sectarian boundaries. Contact, especially that which is not sustained, has been insufficient for transforming dispositions or developing skills of inter-group cooperation (Cairns and Hewstone, 2002). Depth of understanding does not occur through brief encounters with other cultures. Teacher education programs in Southern United States of America and Northern Ireland that do have cross-cultural integration of students in classes also face challenges in accomplishing their goals.

Within teacher education that maintains sustained contact between groups which are divided in their society, development of dispositions, knowledge, and skills are still a challenge. Integrating students of different cultures in the same class is sometimes their first step across a cultural bridge. A climate of openness to all ideas and encouragement for dialogue must be maintained for effective cross-communication. However, common discourse between teachers-in-training is typically guarded and polite to avoid difficult engagement in the conflicts that have historically divided their cultural groups. Their instructors have been challenged with goals of fostering social cohesion between groups and including as topics for classroom discussion the sources of injustice and violence in their societies. Time constraints, especially in one-year training programs, rendered a felt choice between building relationships and positive dispositions for sustaining them or addressing between-group conflicts.

## Examining Cognition

How teacher candidates think about human diversity and conflicts merits investigation in contexts where intercultural relations have been characterized by violence. Analysis of how teacher candidates in two nations described in their narratives major social conflicts revealed cultural patterns in psychological processes that evidenced the need for developing their metacognition (Carter, Koshmanova, and Hapon, 2004). Self-knowledge about one's own thinking and learning is metacognition. Teachers need to be aware of how their thinking habits and knowledge base can affect their instruction. For example, adopting the patterns of dichotomous thinking, such as categorizing disputants in two groups, or characterizing conflicts as interpersonal problems without considering the structural antecedents of them, are methods of analysis that undermine understanding. Analysis skills are not always sufficient for social cohesion and conflict resolution, even when students have reflected on their thought patterns and developed cognitive flexibility to understand from multiple perspectives.

## Preparedness

Teacher preparation for multiculturalism and conflict resolution should positively affect their levels of confidence with those aspects of interaction. With an evaluation of survey and interview data from a cohort of teacher candidates, Hagan, McGlynn, and Wylie (2004) found in Northern Ireland that teachers-in-training of mainly one religious identity felt a need for more experience with diverse populations. Learning in their training program about the need for accommodating human diversity resulted with only 39 percent of the 168 students feeling that their training had sufficiently prepared them. While it is not uncommon for university students to feel their training was insufficient preparation for the extent of practice in their field, efficacy with human diversity is a crucial competency for teachers in territories occupied by -isms. Hence, teacher candidates need preparation that builds their level of confidence with multiculturalism and conflicts associated with it. Evaluation studies of the EMU program evidenced a lack of confidence by classroom teachers for including sensitive and challenging topics in their instruction (Smith and Robinson, 1996). A major criticism of the statutory EMU program was lack of teacher preparation for its use. Another frustration documented in the aforementioned EMU evaluation was the failure of that program for preparing teachers to address in practice social, cultural and political issues in their society. Examination of teacher preparation that includes breadth of experience and depth of involvement with multicultural-based conflicts is worthwhile. Approaches to accomplishing that in a Florida university merit consideration.

## Mandated Diversity Education

The history of intercultural strife in Southern United States of America and the failure of its schools to provide equal education opportunities for all students resulted in statutory courses for teacher preparation, among other mandates. For example, Teaching Diverse Populations (TDP) was a required course for all undergraduate students in Florida who sought teaching credentials. With its foci of pluralism and cross-cultural competencies, the TDP course was a sustained foundation of peace education. Requirements for student deconstruction of their identities and conflict perspectives was one method that enhanced efficacy of teacher candidates in their planning for and practice with peaceful conflict resolution (Carter, 2003). In different reflection essays, students identified as worthwhile their training that extended their learning beyond awareness of diversity and antecedents of associated conflicts to practice with resolving them (Carter and Smith, 2004). Open coding of student's classroom discourse and their reflective writings across several assignments, as well as their responses to anonymous surveys at the end of their courses, facilitated analysis of their training outcomes. One activity that students identified as helpful included their field work in the community where they crossed cultural borders to work with people of diverse backgrounds. Another training strategy that they found to be challenging and transformative for them as well as others was the requirement to recognize social injustice occurring around them in the community, and take action to address it (Carter, 2002b). A precursor to that assignment that helped many students was reflective writing in which they identified their personal and cultural privileges they have. Students also found the practice in the beginning of the course with analyzing diversity-based conflicts, and planning proactive methods of responding to them, expanded their awareness and preparation for peace development in the community. From analysis of data collected during four years at one university's TDP courses, it became evident that students needed experiences with handling diversity issues during their courses (Carter, 2002c). Students also pointed out in their writings a need for role models in their schools and their communities, who are actively working with diverse populations or disrupting social injustice.

## Diverse Perspectives

In Social Studies Methods courses of Southern United States of America, teachers in training learned proactive responses to current societal issues that prepared them to model democratic citizenship. Their writing about the societal issues that they selected and analyzed from a minimum of three perspectives and then took action to address evidenced the value of integrating experience with multiple-perspective analysis before responding to conflicts. Surveys given to teacher candidates for anonymous completion after their citizenship-action assignments revealed their greater capability to understand conflicts from diverse perspectives. Additionally, improved self-conception as proactive role models for teaching their future students was evident in their surveys (Carter, 2002b).

## Summary

Limited instructional standards and cross-cultural integration as well as statutory courses that included foundations of peace were sources of peace-building education for teacher candidates in United States of America and Northern Ireland. How courses with peace foci were taught varied. Cross-cultural peace education in teacher preparation ranged from related-topic inclusion to conflict-transformation experiences, which figure 17.1 illustrates.

Standards for peace-oriented teacher preparation and social education mainly focused on human diversity and democratic responsibilities. However, practice did not evidence widespread use of recommendations beyond those that a university must evidence to maintain accreditation status. Where instruction entailed reflection on outcomes of proactive skill development, teachers in training and their instructors described the value of that integrated instruction across different types of courses. Unmet needs in teacher training for peace building should be addressed by all organizations that influence its practice.

## Needs in Teacher Preparation for Peace Building

Examination of peace-oriented programs for the education of teachers has revealed opportunities for enhancing their outcomes. First, policy and corresponding guidelines determine the enactment of education in United States of America and Northern Ireland (Carter, 2004a). Needed are policies that specifically prescribe educational practice as a foundation for social justice and peace building

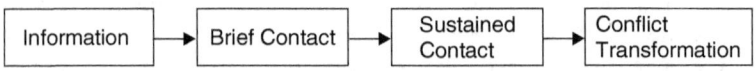

**Figure 17.1**   Continuum of Education Approaches to Cross-Cultural Peace Building

(Carter, 2004b; Darling-Hammond, French, and Garcia-Lopez, 2002). Integrating in a school mission statement the society's goals, such as peace building in schools, helps faculty and students enact that vision. Education standards and criteria for evaluation of teacher education programs must provide specific statements to support peace-building practices in schools. Application of peace-building skills should be recommended because confidence with using them occurs through experience as a component of training.

Second, teacher educators need opportunities for peace-building preparation. They should have their own experiences with development of the dispositions, knowledge, and skills that support effective preparation of their students (Howard, 1999).

Teacher education for diversity involves much more than the transfer of information from teacher educators to their students. It involves the profound transformation of people and of the worldviews and assumptions that they have carried with them for their entire lives (Melnick and Zeichner, 1997, p. 33). For this purpose, universities could provide them time and a place for working on the crucial capacities for peace education. Assisting them on campus together would be more effective than providing instruction on an individual basis for their teacher educators.

Finally, there is a need for temporal opportunities while teacher educators face the issue of sufficient time to include the multiple components of peace building in their curriculum and instruction. Teacher preparation programs that last only two or less years are very challenged with fitting in all that they must accomplish. Integration across teacher training courses of peace components and corresponding competencies for students can be helpful in two ways. First, instructional time is more efficient and course content is richer when curriculum integration is done well. Second, students glean the importance of the integrated theme by encountering it in all of their studies. Teacher candidates should learn within their classes and field practice how integration of peace themes and skills are done in explicit as well as implicit lessons. Peace education could be widespread and sustained when policy, its enactment, and teachers as well as their educators are prepared with experience in peace building.

## References

Anti-Defamation League. (2006, October). Retrieved October 10, 2006 from www.adl.org.
Apple, M. (1995, 2nd ed.). *Education and power*. New York: Routledge.
Banks, J.A. (1997). *Educating citizens in a multicultural society*. New York: Teachers College Press.
Boulding, E. (1988). *Building a global civic culture. Education for an interdependent world*. New York: Teachers College Press.
Bourdieu, P. (1977). Cultural reproduction and social reproduction. In J. Karabel and A.H. Halsey (Eds.), *Power and ideology in education* (pp. 487–511). New York: Oxford University Press.
Brameld, T. (1956). *Toward a reconstructed philosophy of education*. New York: Dryden.
Byrne, S. (1997). *Growing up in a divided society. The influence of conflict on Belfast schoolchildren*. London: Fairleigh Dickinson University Press.
Cairns, E. and Hewstone, M. (2002). Northern Ireland: The impact of peacemaking in Northern Ireland on intergroup behavior. In G. Salomon and B. Nevo (Eds.), *Peace education: The concept, principles, and practices around the world* (pp. 217–228). Mahwah, NJ: Lawrence Earlbaum.

Carter, C.C. (2002a). Conflict resolution at school: Designed for construction of a compassionate community. *The Journal of Social Alternatives*, 21(01): 49–55.

———. (2002b, April). *Preparation of teacher candidates through required social action.* Paper presented at annual meeting of the American Educational Research Association, New Orleans, LA.

———. (2002c, July). *Teacher education for peace: Identifying and addressing seeds of diversity conflicts.* Paper presented at Biannual meeting of the International Peace Research Association, Suwon, South Korea.

———. (2003a, June). *Perspective extensions for cross-cultural conflict resolution.* Paper presented at the UNESCO Conference on Teaching and Learning for Intercultural Understanding, Human Rights and a Culture of Peace, Jyväskylä, Finland.

———. (2004a). Education for peace in Northern Ireland and the USA. *Theory and research in social education, special edition "Peace-building Citizenship Education,"* 32 (1): 24–38.

———. (2004b, April). *Proposed standards for peace education.* Symposium and paper presented at the annual meeting of the American Educational Research Association, San Diego, CA.

———. (2004c). Whither social studies? In pockets of peace at school. *Journal of Peace Education*, 1 (1): 77–87.

Carter, C.C. and Smith, A. (2004, July). *Preparation of teachers for proactive citizenship education.* Paper presented at the biannual meeting of the International Peace Research Association, Sopron, Hungary.

Carter, C.C., Koshmanova, T., and Hapon, N. (2004, April). *Teacher candidates' narratives about extreme social events: Implications for teacher education.* Paper presented at the annual meeting of the American Educational Research Association, San Diego, CA.

Council for Education in World Citizenship, Northern Ireland. (2006). Retrieved October 10, 2006 from http://www.cewcni.org.uk/.

Counts, G.S. (1932). *Dare the schools build a new social order?* New York: John Day.

Darling-Hammond, L, French, J., and Garcia-Lopez, S. (Eds.). (2002). *Learning to teach for social justice.* New York: Teachers College Press.

Dewey, J. (1916). *Democracy and education.* New York: Macmillan.

Educators for Social Responsibility. (2006, October). Retrieved October 10, 2006 from http://www.esrnational.org/.

Eisler, R. and Miller, R. (Eds.). (2004). *Educating for a culture of peace.* New York: Heinemann.

Finn, P.J. (1999). *Literacy with an attitude: Educating working-class children in their own self-interest.* New York: State University of New York Press.

Freire, P. (1998). *Pedagogy of freedom. Ethics, democracy, and civic courage* (Trans. P. Clarke). New York: Rowman & Littlefield.

Fulton, S. and Gallagher, A. (1996). Teacher training and community relations in Northern Ireland. In Craft, M. (Ed.), *Teacher education in plural societies: An international review* (pp. 82–94). Washington, DC: Falmer.

Grant, C.A. and Sleeter, C.E. (1998, 2nd ed.). *Turning on learning: Five approaches for multicultural teaching plans for race, class, gender, and disability.* Upper Saddle River, NJ: Merrill.

Greene, M. (1973). *Teacher as stranger: Educational philosophy for the modern age.* Belmont, CA: Wadsworth.

Hacker, A.H. (1992). *Two nations: Black and white, separate, hostile, unequal.* New York: Scribner's.

Hagan, M. and McGlynn, C.W. (2003, September). Moving barriers: Promoting learning for diversity in initial teacher education. *Journal of Intercultural Education*, 15 (4): 243–252.

Hainsworth, P. (1998). *Divided society. Ethnic minorities and racism in Northern Ireland.* London: Pluto.

Hollins, E.R. (1997). Directed inquiry in preservice teacher education. A developmental model. In King, J.I., Hollins, E.R., and Hayman, W.C. (Eds.), *Preparing teachers for cultural diversity* (pp. 97–110). New York: Teachers College Press.

Howard, G. (1999). *We can't teach what we don't know: White teachers, multiracial schools*. New York: Teachers College Press.
Hudak, G. and Kihn, P. (Eds.). (2001). *Labeling, pedagogy and politics*. New York: Routledge.
Kapur, R. (2004). *The troubled mind of Northern Ireland. An analysis of the emotional effects of the troubles*. London: Karnac.
Koshmanova, T., Carter, C.C., and Hapon, N. (2004). Crisis-response discourse of prospective teachers. *Academic Exchange Quarterly*, 7 (4): 250–255.
Lantieri, L. and Patti, J. (1996). The road to peace in our schools. *Educational Leadership*, 54: 28–31.
Lareau, A. and Horvat, R.M. (1989). Moments of social inclusion and exclusion: Race, class, and cultural capital in family-school relations. *Sociology of Education*, 72 (1): 37–53.
McGlynn, C. (2007). Challenges in integrated education in Northern Ireland. In Zvi Bekerman and Claire McGlynn (Eds.), *Addressing ethnic conflict through peace education* (pp. 77–90). New York: Palgrave Macmillan.
McIntosh, P. (1989). White privilege: Unpacking the invisible knapsack. *Peace and Freedom*, 49 (July/August): 10–12.
McKittrick, D. (2002). *Making sense of the troubles: The story of the conflict in Northern Ireland*. Chicago: New Amsterdam Books.
McLaren, P. (2003, 4th ed.). *Life in schools: An introduction to critical pedagogy in the foundations of education*. New York: Longman.
Mehan, H., Villanueva, I., Hubbard, L., and Lintz, A. (1996). *Constructing school success: The consequences of untracking low achieving students*. New York: Cambridge University.
Melnick, S.L. and Zeichner, K.M. (1997). Enhancing the capacity of teacher education institutions to address diversity issues. In J.I. King, E.R. Hollins, and W.C. Hayman (Eds.), *Preparing teachers for cultural diversity* (pp. 23–39). New York: Teachers College Press.
Morrison, M.L. (2005). *Elise Boulding. A life in the cause of peace*. Jefferson, NC: McFarland.
National Association for Multicultural Education. (2006, October). Criteria for evaluating state curriculum standards. Retrieved October 10, 2006 from http://www.nameorg.org/resolutions/statecurr.html.
National Board for Professional Teaching Standards. (2006, October). Retrieved October 10, 2006 from http://www.nbpts.org.
National Council for Accreditation of Teacher Education. (2006, October). Retrieved October 10, 2006 from http://www.ncate.org.
National Council for the Social Studies. (2000). *National standards for social studies teachers. Volume II, Program standards for the initial preparation of social studies teachers*. Washington, DC. Retrieved October 10, 2006 from http://www.ncss.org.
Noddings, N. (Ed.). (2005). *Educating citizens for global awareness*. New York: Teachers College Press.
Ortiz, P. (2005). *Emancipation betrayed. The hidden history of black organizing and white violence in Florida from reconstruction to the bloody election of 1920*. Berkeley, CA: University of California Press.
Ozmon, H.A. and Craver. S.M. (2003). *Philosophical foundations of education*. Upper Saddle River, NJ : Merrill.
Reardon, B.A. (2001). *Education for a culture of peace in a gender perspective*. Paris: United Nationals Educational, Scientific and Cultural Organization.
Rosenberg, M. (2000). *Nonviolent communication: A language of compassion*. Del Mar, CA: PuddleDancer.
Salomon, G. and Nevo, B. (2002). *Peace education. The concept, principles, and practices around the world*. Mahwah, NJ: Lawrence Erlbaum.
School Mediation Associates. (2005, July). Retrieved October 10, 2006 from *http://www.schoolmediation.com/services/index.html*.
Sergiovanni, T.J. (1994). *Building community in schools*. San Francisco: Jossey-Bass.

Smith A. and Robinson, A. (1996). *Education for mutual understanding. The initial statutory years.* Coleraine, Northern Ireland: Centre for the Study of Conflict, University of Ulster, Coleraine.

Smylie, M.A., Bay, M., and Tozer, S.E. (1999). Preparing teachers as agents of change. In Griffin, G.G. (Ed.), *The education of teachers. Ninety-eight yearbook of the National Society for the Study of Education,* Part I (pp. 29–62). Chicago, IL: University of Chicago.

Takaki, R. (1993). *A different mirror. A history of multicultural America.* Boston: Little, Brown, and Company.

The Collaborative for Academic, Social, and Emotional Learning. (2006, October). Retrieved October 10, 2006 from www.CASEL.org.

TRANSCEND. (2006). A peace development network for conflict transformation by peaceful means. Retrieved October 10, 2006 from http://www.transcend.org.

# Index

Academic service-learning  231, 242
Acculturation  101, 103, 105
Action research  171, 187–190, 194, 197, 199
Activity theory  170, 235
Adult Education  4, 185, 187, 190, 192, 194, 195, 197–199, 201, 202, 204, 205, 208
Alternatives to Violence Project (AVP)  191, 193, 194
Anxiety  51, 78, 88, 101, 123, 127–131, 143, 175
Apartheid  9, 13, 16, 49–51, 53, 56, 63, 71–75, 187, 188, 197
Arabic  93–98, 105, 110, 112–115, 118, 120, 122, 126, 132
Arabs  91, 94, 104, 105, 109, 111, 113, 115, 117, 121, 122, 125, 126, 129, 130
Authoritarian  50, 94, 139, 140, 146, 158, 234
   *See also* Worldview

Bakhtin, M.  235
Balkans  145, 156, 184, 208
Banja Luka  145, 148, 149
Banking model  237
Bekerman  1–4, 46, 55, 58, 91, 92, 94–96, 98, 104, 107, 108, 120, 126, 132
Bernstein, B.  170
Berry, J.  104
Bibler, V. S.  235, 236, 241
BiH Government, *see* Government of BiH
bilingual education  95, 96, 98, 101, 103–105, 107, 109, 110, 113, 115–120
Bilingualism  97, 107, 119, 122, 124, 126

Bosnia and Herzegovina, BiH  138, 139, 144, 145, 147, 148, 150, 153, 154, 156–159
Bosniak  4, 138, 139, 142, 145
Bosnian  139, 148, 151

Catholic  23, 24, 32, 33, 77, 78, 82, 83, 88, 121–124, 127, 128, 130–132, 138, 142, 145, 161, 162, 202–204, 206–208, 212, 215, 232, 251
Christian  18, 58, 93, 99, 138, 142, 162, 174, 199, 210, 232
Citizenship education  27, 28, 30, 32, 33, 75, 239, 256
Coexistence  2, 21, 33, 39, 42, 45–47, 82, 86, 101, 102, 104, 105, 111, 117, 118, 122, 206, 207, 221
Co-existence  88
Columbia University  19, 153, 184
Communities of practice  4, 171, 187, 188, 194–197, 199
Community of practice / or learning  163, 169, 190, 193, 195, 196
Community Peace Program (CPP)  193, 194
Conflict  1, 2, 3, 21–23, 25–27, 29–33, 35–48, 52, 54, 56, 77–80, 82, 84, 86–88, 91–94, 100–105, 107–109, 112, 113, 115–121, 128, 130–133, 137–146, 148, 151–153, 155, 157–159, 162, 163, 166, 167, 170, 173–175, 177–184, 188, 190–192, 198, 201–203, 205–209, 211, 212, 216, 217, 224, 227–229, 231, 232, 245–248, 250, 252–258
Conflict Resolution  1, 26, 29, 30, 32, 36, 39–41, 43, 44–48, 93, 104, 120, 132, 145, 152, 155, 158, 166, 208, 209, 211, 246, 247, 252, 253, 256

## 260 / INDEX

Contact   4, 22, 23, 26, 32, 39, 41, 44, 45, 47, 52, 64, 77, 78, 80, 86–88, 101–103, 105, 108, 113, 118–120, 123, 126, 149, 157, 159, 161, 163, 190, 205–207, 212, 240, 251, 252, 254
Contact theory   4, 105, 133, 157, 205, 207, 212
Control, hierarchical   27, 54, 63, 73, 77, 80, 104, 112, 120, 140, 143, 163, 164, 183, 232, 233, 240
Crime   156, 159, 187, 188
Critical thinking   112, 192, 221, 236, 237, 250
Croat   4, 138, 142, 144, 145, 205
Cross-community contact   26, 39
Culture   3, 9, 21–23, 26, 28–30, 36–45, 47, 50–52, 55–59, 63, 65, 67, 69, 75, 94, 98, 99, 101, 103–105, 108, 111, 113, 117–119, 122, 126, 132, 137–142, 144, 145, 150, 153–155, 157, 169–171, 173–180, 182–184, 188, 196, 198, 203, 205, 207, 208, 211, 219, 222, 228, 229, 231–233, 235, 238–242, 247, 248, 251, 252, 255–257
Culture of Healing, *See* International Education for Peace Institute
Culture of Peace, *See* International Education for Peace Institute
Curriculum EFP   137–158, 184
Curriculum/curricula   1–5, 18, 22, 24, 26–30, 32, 37, 39, 42–45, 52, 53, 56, 58, 71, 74, 77, 79, 83, 85–88, 94, 95, 98, 99, 107–109, 112–114, 118, 119, 127, 129, 131, 135, 138–142, 145, 146, 148, 150, 152–158, 162, 165, 169, 171, 175–179, 181–184, 188–190, 192, 194, 195, 197, 198, 204, 216, 219, 220, 227, 229, 232–234, 237, 240, 245, 246, 248, 249, 255, 257
Cyprus   3, 21–26, 28, 33, 35, 37–48

Danesh, H.B.   4, 137, 146, 148, 156–158
Dayton Peace Accord   138, 139, 158
Democracy   1, 17, 27, 42, 43, 46, 51–53, 56, 58, 71, 73, 74, 93, 105, 147, 157, 187, 189, 197, 199, 219, 233, 235, 241, 256
Development   2–5, 16, 18, 21, 22, 26, 27, 30, 32, 38, 40, 42, 48, 50, 52–54, 63, 64, 79, 80, 82–85, 87, 91–93, 95, 101, 102, 104, 107, 110, 120, 121, 125, 126, 131, 137, 139, 140, 142, 143, 145, 147, 150, 151, 153, 154, 156–159, 162–165, 170, 171, 175, 179, 187, 189, 190–192, 194–197, 199, 201, 204, 206, 207, 209–212, 219–221, 228, 229, 231, 233, 234, 236, 241, 242, 245, 246, 248–250, 252, 253–256, 258
Difference   3–5, 30, 31, 37, 42, 43, 46, 49, 52–56, 66–69, 80, 82, 83, 85, 86, 92–95, 114, 117, 122, 126–131, 138, 142, 147, 166, 168, 170, 172, 173, 179, 180, 182, 208, 210, 216, 218, 219, 227, 238, 247, 248, 250, 251
Diversity   1, 4, 9, 15–18, 27, 32, 42, 50, 52, 54, 56, 57, 59, 73, 75, 77, 78, 81–84, 86–88, 108, 138, 141, 144, 154, 155, 161, 165–168, 173, 174, 192, 207, 208, 211, 218, 222, 227, 236, 240, 241, 248–257
Divided societies   21, 24, 29, 43, 86
   *See also* Segregation
Dunn, S.   33, 161, 162, 165, 170, 171, 206, 211, 212

Education for Mutual Understanding (EMU)   26, 30, 39, 162, 251, 253
Education for Peace   2, 4, 40, 52, 58, 87, 107, 137–139, 141–143, 145, 147, 148, 151–154, 157, 158
Education, multicultural, *see* multicultural education
Educational systems   22, 24, 30, 46
Equality   10, 12, 18, 19, 27, 29, 32, 43, 49, 50, 51–54, 56–59, 75, 78, 83, 87, 92, 95–97, 105, 117, 124, 126, 141, 142, 144, 148, 155, 157, 182, 188, 203, 204, 211
Ethics   211, 223, 224, 228, 229, 256
Ethnocentrism   33, 234, 236, 237, 245
European integration   236, 242
Experiential learning   176, 192, 193, 239

Fraser, G.   33, 88, 171
Freedom   45, 49, 51, 54–57, 102, 110, 111, 113, 141, 144, 155, 157, 187, 198, 256, 257
Freire, P.   198, 228, 242, 256

Gallagher, A.  3, 9, 18, 24, 25, 32, 33, 77, 80, 81, 84, 85, 87, 88, 132, 165, 171, 250, 256
Gellner  91, 105
Generalization  123, 129, 179, 209
Generating affective ties  123, 130
Group  1, 3, 4, 9, 17, 18, 25, 29, 32, 33, 37–39, 41–43, 47–49, 51, 52, 55–57, 63, 64, 66, 70, 71, 74, 77, 78, 82, 83, 85, 87, 93–105, 107, 108, 111–114, 116–120, 122–133, 138–141, 143–146, 149, 153, 158, 159, 165, 170, 173–176, 179–183, 190, 194–196, 202, 205–209, 212, 215, 217, 222–228, 231, 232, 237–239, 247, 248, 251, 252, 255

Haberman, M.  240, 242
Harris, I.  39, 40, 47, 127, 141, 158, 187, 192, 198
Hebrew  93, 98, 108, 112, 114, 115, 118, 122, 126, 132
Hewstone  77, 87, 88, 102, 103, 105, 108, 120, 123, 132, 133, 252, 255
HIV/AIDS  53, 187, 188
Human Rights  27, 28, 30–32, 40, 43, 49, 50, 52, 57–59, 65, 68, 71, 75, 145, 154–157, 159, 196, 228, 256

Identity  9, 22, 23, 24, 25, 27, 29, 30, 31, 32, 33, 41, 42, 47, 48, 64, 65, 69, 70, 72, 73, 74, 75, 77, 78, 88, 92, 93, 95, 96, 99, 100, 102, 103, 104, 105, 109, 112, 119, 122, 126–130, 132, 138, 140, 141, 142, 146, 158, 159, 163, 165, 170, 194–196, 199, 203, 205, 207, 210, 211, 222, 223, 227, 228, 231, 233, 234, 241, 252
  single-identity  22–24, 29, 32, 165, 207, 210
  national identity  22, 24, 27, 30, 31, 33, 95, 99, 128, 129, 132, 231, 233
Individuals  11, 25, 39, 40, 41, 45, 49, 51, 55, 56, 79, 80, 93, 102, 103, 123, 129, 131, 140, 144, 146, 173, 175, 177, 207, 217, 219, 221, 247
Informal learning  166, 171, 197, 199
In-group  98, 123, 130

Integrated Education  2, 3, 24, 25, 33, 77, 78, 79, 80, 81, 82, 83, 84, 85, 86, 87, 88, 89, 93, 95, 97, 99, 101, 103, 104, 105, 107, 122, 124, 128, 131, 161, 162, 163, 164, 165, 166, 167, 168, 169, 170, 171, 202
Integrated Schools  3, 4, 13, 23, 24, 25, 30, 33, 39, 77, 78, 79, 80, 81, 83, 85, 86, 88, 95, 99, 117, 121, 122, 127, 128, 129, 130, 131, 132, 161, 162, 163, 164, 165, 166, 168, 169, 171, 246
Integration  2, 3, 14–17, 24–26, 32, 52, 54, 59, 64, 68, 70, 74, 75, 77, 81, 82–88, 96, 103, 105, 108, 133, 165–169, 177, 210, 221, 234, 236, 242, 251, 252, 254, 255
Integrative  37, 41, 78, 132, 137, 141, 148, 153, 154, 156, 157, 159, 176, 182–184, 193
Integrative Theory of Peace, ITP  137, 139, 154, 156, 157
Interethnic conflict  21, 22, 232
Interethnic reconciliation  149
Inter-group  248, 252
Israel  1, 3–5, 18, 40, 47, 91–112, 114–132, 159, 184, 202

Jew  123
Johnson, L.S.  3, 21, 22, 24, 25, 29, 30, 32, 33, 162, 165–168, 171, 192, 199
Justice  12, 16, 18, 21, 22, 27, 29, 30, 31, 51, 53, 56, 57, 68, 71, 92, 141, 143, 144, 146, 148, 155, 168, 170, 182, 189, 192, 199, 206, 211, 216, 233, 235, 239, 247–250, 252–254, 256

Knowledge  5, 22, 26, 27, 28, 29, 30, 43, 52, 55, 84, 99, 101, 107, 114, 115, 116, 119, 122, 123, 125, 141, 157, 163, 164, 165, 168, 169, 170, 171, 175, 176, 177, 179, 182, 183, 194, 195, 196, 197, 199, 204, 205, 206, 210, 219, 220, 221, 223, 224, 233, 237, 239, 243, 245, 246, 247, 248, 249, 250, 251, 252, 255
KwaZulu-Natal, South Africa  4, 157

Language  9, 23, 25, 31, 37, 43, 45, 51, 52, 53, 54, 55, 57, 58, 64, 67, 74, 81, 93, 94, 95, 97, 98, 103, 105, 107, 108, 110, 112, 114, 115, 117, 118, 122, 126, 162, 163, 169, 174, 192, 202, 204, 207, 208, 222, 224, 232, 233, 242, 257
Lave, J.  161, 163, 164, 166, 169–171, 195, 196, 199
Learning
  community of  169
  informal  166, 171, 197, 199
  theories of  195
Love  125, 129, 142, 148, 176, 238, 241

Marshal, Tito  138
McGlynn, C.  1, 3, 25, 33, 46, 77, 78–84, 86–88, 165, 171, 246, 252, 256
Mediator  123, 125, 130, 205
Mentoring  195
Meta-analysis  123
Mezirow, J.  193, 199
Minority rights  9, 18, 170
Moderator  123
Moslem  37, 93, 99
Multicultural education  1, 59, 74, 75, 87, 88, 100, 104, 161, 162, 170, 235, 237, 242, 249, 257
Multiculturalism  30, 40, 42, 45, 46, 71, 87, 88, 96, 100, 101, 163, 170, 233, 235, 237, 249, 252, 253

National identity, see identity
Neutral  127, 183, 202, 203, 206
Northern Ireland  1, 3, 4, 5, 18, 21, 23- 33, 47, 77–87, 89, 90, 123–126, 129–135, 159, 163–167, 170, 172, 173, 175, 204–215, 248–251, 254, 255, 257–258

Office of High Representative, OHR  138, 147, 157–159
Organization for Security and Co-operation in Europe (OSCE)  156
Out-group  25, 123, 125, 126, 130

Palestinian  3, 47, 58, 91, 92, 93, 94, 95, 96, 97, 98, 99, 100, 101, 103, 104, 106, 107, 109, 110, 111, 112, 113, 114, 115, 116, 117, 118, 120, 121, 130, 133, 160, 203

Partnership  31, 44, 84, 193, 243
Peace education, see education for peace
Peace event  142–145, 156
Peace moves  158
Peace-based curriculum  139, 155
Pedagogic practice  163
Pedagogical Institutes  150, 157
Pedagogy  4, 22, 28, 30, 32, 58, 59, 74, 95, 104, 110, 119, 154, 161, 163, 164, 165, 166, 168, 169, 170, 171, 189, 193, 197, 226, 227, 228, 234, 239, 240, 241, 242, 246, 247, 249, 257
  rights-based  28
Pedagogy of Civilization  154
Pedagogy, visible, invisible  4, 163, 164, 166, 167, 169
Pettigrew and Tropp  123
Poverty  52, 143, 240, 242
Prejudice  1, 2, 22, 31, 44, 52, 70, 78, 87, 88, 101, 102, 104, 105, 108, 116, 120, 123, 129, 130–133, 141, 144, 159, 161, 163, 165, 167, 204, 205, 210, 211, 217, 220, 221, 224, 226–228, 236, 239
Principles of Peace  137, 142, 143, 147–149, 152–154, 156
Process-oriented, see Curriculum
Protestants  23, 24, 32, 77, 78, 82, 83, 88, 121–124, 127, 128, 130–132, 161, 162, 202–204, 206, 207, 212, 251

Reflective practice  164, 171, 172, 188
Religion  24, 51, 71, 98, 179, 180, 202
  Catholic  23, 24, 33, 77, 78, 82, 83, 122, 123, 127, 128, 132, 138, 142, 145, 162, 202, 203, 207, 208, 212, 215, 251
  Muslim  210
  Orthodox  93, 94, 138, 232
Religious balance  162, 166

Sarajevo  139, 143, 145, 148, 149, 158
Sectarianism  1, 33, 79, 87, 128, 203, 206, 245
Self-awareness  164
Self-disclosure  125, 130
Serb  4, 138, 139
Service Learning  5, 231, 236, 238, 241, 242

Social Cohesion   4, 21, 22, 24, 31, 32, 78, 86, 229, 231
Social Justice   30, 52, 53, 71, 211, 239, 254
South Africa   3, 4, 9, 49–59, 63–65, 67, 68, 71–75, 157, 184, 187–189, 194–199, 201, 208
Spirit   50, 67, 73, 83, 166, 221, 231
Spirituality   158
Srebrenica   139
Stereotypes   29, 39, 44, 47, 87, 105, 111, 116, 123, 125, 126, 130, 132, 163, 205, 216, 220–228, 234, 236–239, 241
Sustainability   22, 79, 101, 152, 153, 193
Swiss Development and Cooperation Agency, SDC   150, 151, 156

Teacher training   5, 22, 24, 27, 30, 71, 117, 131, 228, 234, 245, 248, 250, 251, 254, 255
Teacher Unions   22
Threat   27, 42, 92, 123, 128, 130, 131
tolerance   22, 30, 49, 50, 52, 56, 59, 78, 110, 111, 116, 124, 125, 126, 127, 129, 130, 169, 179, 183, 204, 205, 221, 225, 231, 235, 236, 237, 241, 242
Transformative learning   193, 199
Travnik   142, 144, 145, 148, 149
Truth and Reconciliation Commission, TRC   50, 57, 144

Ukrainian nationalism   234, 242, 243
Understanding-oriented, *see* Curriculum
Unemployment   174, 187, 188

United Nations   23, 38, 147, 157–159, 172, 173, 175, 184, 201, 211, 212
United States Institute for Peace, USIP   153, 156
Unity   50, 87, 138, 139, 140, 141, 142, 143, 144, 146, 149, 151, 152, 153, 154, 155, 156, 183, 217, 219, 221, 234
Universal   55, 69, 78, 100, 137, 138, 142–144, 153, 156, 221
University of KwaZulu-Natal, South Africa   157

Violence (domestic, political, culture of)   1, 21, 38–42, 46, 49–52, 91, 117, 137–142, 144, 147, 152, 153, 155, 157–159, 173–176, 179, 180, 182–184, 187, 188, 190–193, 198, 199, 207, 211, 215, 216, 218, 220, 235, 245–250, 252, 257
Violence-Free Schools   153

Web-based Education for Peace, *see* EFP
Wells, G.   19, 163, 170, 172
Wenger, E.   161, 163, 164, 166, 169–171, 195, 196, 199
Worldview   21, 27–29, 42, 43, 67, 69, 71, 72, 125, 137–143, 146, 148, 152, 154–157, 221, 223, 255
  Survival-Based   138, 140, 154
  Identity-Based   138, 140, 154
  Unity-Based   138–143, 146, 152, 154–156
  Peace-based   139, 140
  Attitude Change   148, 221

Youth Peacebuilders Network (YPN)   156
Yugoslavia   1, 138, 139, 205, 231

GPSR Compliance

The European Union's (EU) General Product Safety Regulation (GPSR) is a set of rules that requires consumer products to be safe and our obligations to ensure this.

If you have any concerns about our products, you can contact us on

ProductSafety@springernature.com

In case Publisher is established outside the EU, the EU authorized representative is:

Springer Nature Customer Service Center GmbH
Europaplatz 3
69115 Heidelberg, Germany

www.ingramcontent.com/pod-product-compliance
Lightning Source LLC
LaVergne TN
LVHW011809060526
838200LV00053B/3709